PLANT IMMUNITY
Biochemical Aspects of Plant Resistance
to Parasitic Fungi

PLANT IMMUNITY
Biochemical Aspects of Plant Resistance to Parasitic Fungi

By L. V. METLITSKII and O. L. OZERETSKOVSKAYA
A. N. Bakh Institute of Biochemistry
Academy of Sciences of the USSR

Translated from Russian

 SPRINGER SCIENCE+BUSINESS MEDIA, LLC 1968

Born in 1913 and graduated from K. A. Timiryazev Agricultural Academy in Moscow, Lev Vladimirovich Metlitskii holds a doctorate in agricultural sciences and is presently director of the Laboratory of Biochemistry of Plant Immunity at the A. N. Bakh Institute of Biochemistry of the Academy of Sciences of the USSR. He has published over 100 articles, mostly on the biochemical nature of resistance and passivity in plants and the biochemical principles of crop immunity to infectious diseases.

Ol'ga Leonidovna Ozeretskovskaya was born in 1932 and is a graduate of Moscow State University. She is presently a senior scientist in Prof. Metlitskii's laboratory.

A Russian edition of this work is to be published by Nauka Press in Moscow in 1968. The translation was prepared from a copy of the manuscript submitted to the Soviet publisher.

Л. В. Метлицкий, О. Л. Озерецковская

ФИТОИММУНИТЕТ
Биохимические аспекты устойчивости растений
к паразитарным грибам

ISBN 978-1-4899-7300-9   ISBN 978-1-4899-7317-7 (eBook)
DOI 10.1007/978-1-4899-7317-7

Library of Congress Catalog Card Number 68-25383

© 1968 Springer Science+Business Media New York
Originally published by Plenum Press in 1968
Softcover reprint of the hardcover 1st edition 1968

Contents

Introduction

The problem of plant immunity to infectious diseases is one of the most interesting and complex of biological questions. An understanding of the internal mechanisms by which plants protect themselves from parasites under natural conditions would facilitate the development of effective means for the protection of cultivated plants from disease.

It is well known that the development of immune varieties is an effective method of guarding plants against disease. But the practice of selection shows that it is much easier to develop resistant than immune varieties, i.e., those relatively slightly affected by disease. The more resistant the variety, the lower the requirement for chemical preparations to eliminate the infection and the more effective their action. Under the conditions of modern agriculture this is of special significance. At the present time the residue of a number of fungicides in agricultural products often exceeds permissible norms. Moreover, of no less importance is the fact that, with the expanding rate of fungicide usage, we inevitably contribute simultaneously to the development of resistance in phytopathological microorganisms, as a result of which the protection of plants from disease becomes more and more complicated.

Currently, investigation of the greatest possible number of prospective substances and detection among them of a compound with high selective action is the only method of searching for effective fungicides. This compound should be lethal to the micro-

1

organism without causing damage to the host plant. It is quite probable that the substances that assist plants to protect themselves from infection and that establish their resistance to many diseases would be a good standard for a fungicide. These substances, as we understand, are either already present in the plants before contact with a pathogen, at which time their content may increase greatly in the process of interaction with the parasite, or they are formed *de novo* in infected tissues. A study of the biochemical nature of phytoimmunity leads to the conclusion that there is a real possibility of formulating preparations that would not only exert a toxic effect on phytopathogenic organisms but would simultaneously stimulate development in the plants themselves of protective substances against parasites. A successful solution to this problem would mark a new, extremely important step in the protection of cultivated plants from infectious disease.

It should nevertheless be emphasized that, with the great significance of plant protection from infectious diseases, the problem of phytoimmunity should not be confined to this alone. Burnet [1962] notes that the broadest concept of immunity appears to be embodied in the role it plays in processes directed toward maintenance of the structural and functional integrity of any organism.

According to the results of Stakman and Harrar [1957], more than 100,000 different fungi parasitize the higher plants, i.e., permanently or temporarily live on their surfaces, within them, or cooperatively with them, obtaining nourishment from them in varying quantity and in different qualitative form. This naturally accounts for the diverse character of interrelationships between plants and parasites and, consequently, for the diverse nature of resistance in plant tissues. Vavilov [1935] has emphasized repeatedly that no single mechanism of resistance to the different causes of infectious disease is likely to exist for all plants without exception.

These concepts notwithstanding, the wide distribution of phytoimmunity in nature and the alteration in character of interrelationships between plant and parasite during the course of evolution automatically suggest that, even with the extreme diversity of plants and parasites, the study of plant immunity must be based on a search for general principles.

The progress made by biological chemistry in recent years permits new approaches to the study of the biochemical nature of plant immunity and the revision of ideas which until quite recently seemed unassailable. An attempt to review within the limits of a short book the results of all investigations would, however, be too difficult a task. In the present volume we shall examine only certain biochemical aspects of the resistance of plants to fungus diseases, basing ourselves frequently on work carried out in our laboratory at the A. N. Bakh Institute of Biochemistry of the USSR.

Chapter I

Problem of Material and Process
in the Phenomena of Phytoimmunity

Hippocrates had already learned that the disease process includes not only the disease as such, i.e., the damage to the organism, but also the struggle of the organism for restoration of the norm. This position is fully admitted by contemporary doctrine regarding diseases, which, along with questions of etiology, attaches special significance to the problems of the mechanisms that control the internal environment and provide for the restoration of its normal state when it has been disturbed [Zdrodovskii, 1961]. The overwhelming role in the maintenance of the structural and functional unity of any organism belongs to immunity. Nevertheless, the definition of phytoimmunity as an expression of the norm of reaction characteristic of a higher plant to interference in its life activity by a heterotrophic organism, or, in other words, the statement that the higher plant can remain healthy only by means of an active reaction to the intrusion of a parasite, is controversial [Rubin and Artsikhovskaya, 1960]. Not only active but also passive immunity is considered in the phytopathological literature.

The results of countless investigations, however, lead one to the conclusion that the division into active and passive plant immunity is often quite arbitrary, a fact already noted by Vavilov [1935]. Frequently the plant reaction is so difficult to observe that a false impression of its complete absence is created. It is for this reason that some authors refer to passive and others, with equal justification, to active immunity. For example, according to

5

Gorlenko [1962], the formation of phytoncides by plants is an element of passive immunity since phytoncides exist in intact tissue. According to Verderevskii et al. [1964], this process is, on the contrary, a typical element of active immunity, since their studies indicate that phytoncide activity of tissues increases appreciably in response to infection. Verderevskii proposes that only such factors in the protection of plants from disease as involve no contact between the pathogen or its toxins and living plant tissues be ascribed to the phenomenon of passive immunity.

Anatomical−morphological characteristics of a plant, including the structure of its integumentary tissues, are often referred to passive immunity. But in the great majority of cases the integumentary tissues are not only a mechanical obstacle in the pathway of infection but also a chemical barrier because of the presence in them of substances possessing antibiotic activity. We shall be convinced of this on analysis of the protective role of plant reactions to wounding.

In any case, the deciding role in protection from infection belongs to active immunity. In this connection, a study of the biochemical nature of the active responsive reaction of plants to infection that determines the final outcome of the interrelationships between host plant and parasite is of prime importance.

Even the initial works dealing with this question showed a tendency to connect the different degrees of plant resistance with peculiarities of the chemical composition of their tissues. These studies, which had already begun at the end of the last century, were based on the different requirements of microorganisms for nutritive substances. Therefore, an attempt was made at first to associate differing resistance to disease with the degree of assimilability by the parasites of the nutrient substances contained in the plants and with their diffusion through the cell walls [Pfeffer, 1884; Miyoshi, 1894; Massee, 1905; Comes, 1914; and others].

Thus originated the chemotropic theory of plant resistance to infectious disease. According to its advocates, positive chemotropism lies at the basis of penetration of parasites into a plant; from this follows that those plants in which substances assuring positive chemotropism are absent should be immune.

This explanation of plant resistance, however, was quickly disproved. In many cases the parasites grew equally well both toward chemotropically positive substances and in the opposite direction. A critical consideration of the chemotropic theory of plant resistance to infectious diseases has been presented by Vavilov [1935] and also, subsequently, by other workers.

This theory is now mainly of historical interest; nevertheless, these studies were an important stage in the course of investigation of the nature of phytoimmunity. Their principal importance lies in the fact that they focused the attention of scientists on a more detailed study of the role of the chemical composition of plants as an index of their resistance.

In many cases plant immunity may be based on the fact that the plant environment does not satisfy certain requirements of microorganisms, i.e., it is deficient for them. This theory is still called the nutrition hypothesis or the theory of deficiency of the environment. It is based on the inability of many fungi and bacteria to synthesize independently certain compounds essential to life: amino acids, vitamins, and nucleotides. From this point of view, the virulence of a parasite is determined by its ability to synthesize the deficient compounds.

. The nutrition theory applies in some measure to the combination plant—obligate parasite, since the obligation itself attests to the inability to synthesize certain compounds, and a high degree of specialization of parasite to plant can make the latter unsusceptible if it loses its ability to supply the parasite with a particular substance. But, in the case of facultative parasites, such a mode of protection is not applicable.

Over thirty years ago, A. I. Oparin first showed experimentally that metabolites of normal exchange inherent to the plant organism can possess antibiotic properties, i.e., inhibit or kill microorganisms. The presence of antibiotic substances, antibiosis, was observed among intermediate products of metabolism in garden-beet roots. And, although the chemical nature of antibiosis of beet roots was not clarified, a number of experiments succeeded in showing that its activity is directly related to the physiological state of the plant itself. Thus, growth inhibition of yeasts was most intense with sap from healthy roots; it was much weaker for frozen and still weaker for withered roots [Oparin and Kuplenskaya, 1935].

Tokin [1942, 1948] first showed that all plants without exception produce antibiotic substances, which he called phytoncides.

The initial studies served as foundation and stimulus for the formulation of new experiments in many laboratories. As a result, it became clear that numerous antibiotic substances, most diverse in chemical nature, are found in a single plant. Unfortunately, this fact is frequently overlooked by workers who question the decisive role of antibiotic substances in phytoimmunity on the ground that in microorganisms the ability to adapt to poisons is strongly developed. In fact, under laboratory conditions, multiple transfer of microorganisms on artificial nutrient medium with a gradual rise in the concentration of poison finally makes possible their growth even at high concentrations of the inhibitors. Such experiments, however, are quite remote from what takes place under natural conditions. The principal difference consists precisely in the fact that in the plant cell not one but several inhibitors are formed. The biological significance of the formation by a plant of a whole series of substances toxic to the parasite lies in their synergistic action by which the toxic effect is greatly increased. Also not excluded is the possibility of the presence in the plant of substances which, while not antimicrobial themselves, enhance the activity of the inhibitors. For example, the biologically inactive detergent Tween-80 increases the effectiveness of streptomycin against tubercle bacteria thousands of times [Lacey, 1958]. The same also applies to a mixture of chlorogenic and isochlorogenic acids [Kirkham, 1957], chlorogenic acid and amino acids, etc.

The presence in plants of certain antibiotic substances that, in addition, act synergistically is apparently one of the reasons why the higher plants, greatly exceeded by microorganisms in rate of multiplication, versatility, variability, and adaptability to unfavorable conditions, nevertheless are not defenseless against parasites, which accounts for the fact that death of (wild) plants in nature is the exception rather than the rule [Metlitskii and Akhvlediani, 1965]. A similar point of view is expressed by Kuč [1966], who also holds that the multiplicity of inhibitors in plant tissues explains the resistance of plants under natural conditions.

It has been shown experimentally that only very few individuals in a test population of microorganisms attain the genetically transmitted ability to withstand the presence of even one poison in the substrate. The probability of the appearance of such an ability in certain bacteria has been determined to lie in the range of 10^{-6}-10^{-8} [Petrov, 1959]. With the presence in the substrate of a second poison, the probability of adaptation is 10^{-12}-10^{-16}; with three poisons, 10^{-18}-10^{-24}; etc. The probability of 10^{-24} means that of 10^{24} individual microorganisms only one can acquire the inheritable ability to be resistant to all three poisons.

It is true, of course, that these data apply to bacteria and not to phytopathogenic fungi and that they were obtained in experiments *in vitro*. Also, plant tissues may contain substances that stimulate the growth of a parasite as well as mutagenic compounds, as a consequence of which the true probability of the appearance of an individual resistant to the antibiotic substances of the plant may prove to be much greater. But this in no way minimizes the effectiveness of a synergistic action of antibiotic substances contained in plants.

The degree of toxicity of any substance can by no means be judged by its total content in a plant. Such substances are most often localized both in individual tissues and within cells. Many of the antibiotic substances possessing protein-denaturing ability cannot, naturally, occur in soluble form in the cell protoplasm. This is true of polyphenols, which are contained primarily either in the cell walls or vacuoles. It is possible that at the time of infection the protective substances are able to pass into the intercellular space, where their concentration becomes very high. Finally, on localization of antimicrobial substances in the surface area, their concentration may be so marked that it is entirely incomparable with the average content in the plant.

Another extremely important fact should be kept in mind here. Antibiotic substances, in the course of conversion, can form a number of compounds that have either very slight antibiotic properties or none whatsoever, or that may even exert a stimulatory effect on the growth of phytopathogenic microorganisms. For example, the contradictory results in the literature concerning the protective role of essential oils in the phenomena of plant resistance to mold fungi are explained by the fact that plant resistance has been related to the total content of essential acids. But only certain products of their enzymatic breakdown possess antibiotic properties, especially allicin, which forms as a result of the decomposition of alliin, a basic component of the essential oils of many plants.

Korableva and Potapova [1966] in our laboratory investigated the relationship between the resistance of onion to *Botrytis allii* and the content and conversion of essential oils. The greatest amounts of essential oils were observed in the strong onion varieties and also in the meristematic tissues of the bulb, which are characterized by a higher resistance. Within the limits of one variety, however, the resistance of the bulbs to fungus infection was, inversely, least at the time of the highest content of essential oils in their tissues.

This seeming contradiction may be explained by the fact that onion resistance is associated not so much with the total content of alliin as with the intensity of its conversion and the formation of allicin, and possibly also of ammonia. In intact tissue the reaction of alliin breakdown is most strongly expressed at the times of greatest active vitality of the bulb, maturation immediately after harvest and spring germination, at which times the bulbs are also characterized by the highest resistance.

Results of this type were obtained in our laboratory by Sokolova et al. [1967] in a study of the resistance of beet roots to storage rot. A direct correlation was observed between the resistance of different parts of the root and the activity of choline dehydrogenase responsible for the conversion of choline to betaine, which possesses definite antibiotic properties. The activity of choline dehydrogenase was at its peak in the root cap, which is also characterized by the highest resistance.

But if antibiotic substances are able to exert a lethal effect on an invading parasite, then how can the plant in which they occur remain healthy? For this purpose, during the prolonged evolutionary process various defense reactions have developed in the plant organism; one of these consists in the death of a portion of the surrounding cells of the host plant which accompanies penetration of the parasite. This is the so-called necrotic reaction, or the reaction of hypersensitivity, through which, by destruction of the parts, the organism as a whole is saved.

It must also be emphasized that adaptation of microorganisms to antibiotic substances of plants and its important role in the development of parasitism cannot be entirely excluded. From the use of antibiotics in medical practice, it is known that some pathogens have already attained resistance to a number of antibiotics. This does not necessarily apply, however, to the adaptation of microorganisms to fungicides and to antibiotic substances of plants, if only by virtue of their multiplicity. It is necessary also to distinguish the adaptation of individual microorganisms associated with the norm of genotype reaction to some external environment (phenotypic adaptation), from the adaptation of the population based on the association of processes involving variability, heredity, and selection (genetic adaptation), with which we are here concerned.

Dubinin [1967] considers that the only source of new genetic characteristics of organisms is mutation, i.e., molecular changes in individual genes and changes in the number or structure of the chromosomes. In addition, he submits many facts indicating that the appearance of mutations depends not only on the properties of the genetic material but also on extra-genic conditions external to the genetic structures.

According to Zhukovskii [1966], the point where an adaptation of the parasite to a host variety is suspected is just where a new race should be sought.

Thus, the questions of phytoimmunity are linked most intimately with those of genetics. Unfortunately, little work on the genetics of plant immunity, and especially on the biochemical aspect of this problem, is found in the literature. Nevertheless, both complete immunity, based on incompatibility of the plant host and the parasite, and the degree of resistance of a plant to a parasite are hereditary characters regulated by the genetic apparatus of organisms [D'yakov, 1965].

Among the significant results of genetic investigations is the observed correlation between the number of genes determining the resistance of a given plant species to a given parasite and the number of races of this parasite. The more obligate (biotrophic) the parasite in method of nutrition, the greater the number of its races and the greater the number of genes controlling the resistance of the host plant. Thus, the genetics of parasite virulence reflects the genetics of plant resistance, and vice versa.

A similar discovery was made by Flor [1956] through the observation in flax of 25 genes for resistance to rust and an equal number of virulence genes in the flax rust pathogen *Melampsora lini* . The system of "gene against gene" discovered by Flor corresponds not only to the interrelationships of flax with rust but also of wheat with stem rust and with hard and powdery smut, of corn with rust, of barley with powdery mildew, of potato with Phytophthora, etc.

Amplifying these studies, Person [1959] worked out methods of analysis for the system of gene to gene. If the number of resistance genes in all varieties of a plant species phenotypically representing resistance is designated as n, then the number of parasite races that can occur for them is 2^n. Thus, the resistance of potato to *Phytophthora infestans* is controlled by four dominant genes designated R_1, R_2, R_3, and R_4. In the presence of four resistance genes, 16 races of Phytophthora would be expected to coordinate with the corresponding host genotypes. The prediction has proved to be correct. At the present time, 19 races of Phytophthora have already been recorded, corresponding to additional resistance genes observed in potato [Malcolmson, 1965].

Within the framework of the present book, we are not able, of course, to examine genetic studies in the area of phytoimmunity. We shall attempt only to demonstrate, with phytophthorosis of potato as the example, how from the genetic point of view plant resistance characteristics can be vanquished by means of the generation of new virulent races.

More than a quarter of a century ago, Müller and Börger [1940] showed experimentally that the virulence of new Phytophthora races cannot explain their adaptation to the potato inhibitors. In the course of their experiments, potato tubers were inoculated with incompatible races of *P. infestans*, in response to which the necrotic reaction appeared, associated with the formation of fungitoxic substances. Thereafter, these same tubers were inoculated with a virulent race of the same parasite. Infection, however, did not take place. It appears that fungitoxic substances developing in response to the inoculation with a nonvirulent race inhibit equally the growth of both the virulent and the nonvirulent races of the parasite. In this connection, the virulence of a compatible race of P. *infestans* cannot be explained by its adaptation to the fungitoxic substances of the host plant. Thus, it is not a matter of adaptation of the parasite to these substances but of inability of the plant host to form them in response to the introduction of the virulent race.

This, in turn, may have two explanations. First, a virulent, as distinguished from an avirulent, race injures that apparatus in the plant cell which responds with the formation of protective substances in the required quantity and with sufficient speed. Second, the virulent, unlike the avirulent, race does not make those substances (inductors) without which the plant host is unable to activate its protective apparatus. In other words, the defense apparatus exists but remains inactive upon introduction of the virulent race [Gallegly and Niederhauser, 1959].

D'yakov and Kogan [1966] repeated the above experiment of Müller and Börger, but in reverse. Potato tubers were first inoculated with a virulent race of P. *infestans* and after 24 hours the same point in the tuber was inoculated with an avirulent race. If the inoculation first of a virulent race destroys the integrity of the protective apparatus of the tuber, then the following inoculation with an avirulent race should not induce the hypersensitivity reaction. But actually the second inoculation gave a typical hypersensitivity reaction: tissue necrosis and strong suppression of sporulation in the parasite.

On the basis of this experiment, D'yakov and Kogan conclude that the virulence of races of P. *infestans* is dependent not on a disturbance of the mechanism for biosynthesis of protective substances within the host plant, but on the noninduction of this mechanism in response to infection. Virulent races of P. *infestans*, unlike the avirulent, can bypass the protective apparatus of the host, which thus appears inactive.

This then suggests that the appearance of a virulent race of a parasite is dependent on the ability of the latter not to produce that inductor in response to which the plant cell can activate its defense mechanism. At least this is represented as one of the possible reasons for the inability of the plant cell to respond with the hypersensitivity reaction to the invasion of a virulent race of the parasite.

With the appearance in microorganisms of the ability to overcome the antibiotic substances and to use the nutritive substances
of plants, they become more dependent on the host plant and conditions of the external environment. Unlike saprophytes, which develop on the same varied organic substrates, specialization of certain parasites to the host plant reaches a point where they become
limited not only to rigidly defined plants, but even to individual or-
gans. Such a severe adaptation of parasites to host plants naturally complicates the struggle against them, but, on the other hand,
removes the threat of parasites with the omnivorous ability of
saprophytes, which man might not be able to control.

Vavilov [1935] considered that the biological specialization
of parasites to a host is the most important factor in plant immunity, and consequently he distinguished generic, species, and
varietal immunity. With generic and species immunity, the most
widespread forms of phytoimmunity, entire genera and species are
unaffected by certain diseases. With varietal immunity, individual
varieties are unaffected within the limits of one species.

In the phenomena of generic and species immunity, noncon-
formity of parasite to host plant is so strongly expressed that complete lack of reaction by the plant to invasion by the parasite is
suggested. But, actually, this is far from the fact. It was noted
long ago that, in a whole series of cases, under favorable conditions (temperature, humidity, etc.) spores of parasites are able to
germinate on the most widely differing plants. Differences in the
degree of their specialization appear later — after penetration of
the parasite into the plant tissues. As we shall see below, antibiotic substances (phytoalexins) develop in response to contact, not
only with specialized parasites, but in still greater degree with un-
specialized ones. Thus, experimental confirmation has been obtained of Vavilov's view that there is no fundamental difference between species, generic, and varietal immunity.

The chemical nature of antibiotic substances of plants will be
treated in the following chapter. We shall mention here only that
all antibiotic substances observed in plants have proved to be relatively low-molecular compounds. The question naturally arises as
to the role of proteins in plant resistance to disease.

Many workers have established that the activity of certain enzymes rises in response to infection; that of others, on the contrary, drops; and all enzymes, as is well known, are proteins. Thus, the protective role of proteins consists primarily in their regulation of those chemical reactions that originate in the plant in response to infection and that are basic to immunity. However, to what degree the rise in enzyme activity observed here is the result of newly formed protein remained unknown for a long time and the answer to this question has been found only with the development of new techniques for protein isolation and purification.

In many cases it has been established that the protein in the plant rises appreciably in response to infection. This includes protein soluble in the cytoplasm and also protein of the cell structures. According to the results of Akazawa [1956] with sweet potato tissues infected with *Ceratostomella fimbriata*, the content of protein nitrogen in the fraction of soluble cytoplasmic protein of the mitochondria and microsomes increased 125-150%.

The fraction of soluble cellular protein-albumins has been studied in the greatest detail. Heitefuss et al. [1960] succeeded in observing that infected leaves of cabbage of a susceptible, in contrast to a resistant, variety contain three new protein components. In other cases, the appearance of new proteins was seen in the protein fractions of resistant varieties. For example, Kawashima and Uritani [1963] observed their development in tissues of a resistant variety of sweet potato infected with *C. fimbriata*. A change in the components of the soluble proteins was noted by Tomiyama and Stahmann [1964] in potato tubers resistant to *P. infestans*. A number of new proteins was observed on infection of beans with the rust pathogen *Uromyces phaseoli* and with the bacterium *Pseudomonas phaseolicola* [Rudolph and Stahmann, 1964].*

---

* In recent times a tendency has been noted to ascribe the rise in enzyme activity in tissues of infected plants solely to protein synthesis. The work of Farkas et al. [1964] warns against excessive enthusiasm in this regard. The authors noted that in detached wheat leaves enzyme activity rose almost the same as in infected leaves. The increased activity in cut leaves can hardly be attributed to the synthesis of enzymatic protein. The authors believe that the reason is more likely the proteolysis of such cell structures as ribosomes and mitochondria, as a result of which the enzymes are liberated from an inactive form. It is not impossible that a similar pic-

A qualitative study of newly formed proteins revealed that many of them were isozymes of enzymes already present in healthy tissue. Isozymes of polyphenoloxidase and peroxidase were observed in large quantities. Thus, in the study of a potato variety resistant to Phytophthora nine isozymes of polyphenoloxidase appeared in tissues adjacent to the point of infection. In cytoplasm from leaves of healthy beans, two isozymes of malate dehydrogenase have been noted, and in leaves infected with rust their number rose to four [Staples and Stahmann, 1963]. The number of isozymes of succinic acid dehydrogenase also rose, as did those of the alkaline and acid phosphatase. In some cases, suppression of one of the isozymes of acid phosphatase of bean infected with *Pseudomonas phaseolicola* was observed [Rudolph and Stahmann, 1964].

The developing isozymes can be distinguished by substrate specificity and degree of resistance to an unfavorable influence. Thus, Nelson and Dawson [1944] divided the polyphenoloxidase of the fungus *Psalliota campestris* into two components, one of which oxidized catechol with great rapidity, and the other cresol. Later, Jermyn and Thomas [1954] separated the peroxidase of horseradish into five components which differed in oxidation substrate.

In the light of the above, it seems most probable that isozymes appearing in plants in response to infection are characterized by a higher resistance to the toxic secretions of the parasites. In defense reactions of plants this would be of utmost significance. The protective role of newly formed isozymes can consist also in intensification of the formation of fungitoxic substances.

New techniques in the study of plant proteins have reawakened interest in the old question of the possibility of antibody formation in plants, just as in animal organisms. Nevertheless, no new evidence as to the existence of such a defense mechanism in plants has yet been obtained. Moreover, it is unlikely, if only because of the lack of a circulatory system by which antibodies in the animal

---

ture may be found in infected plant tissues, as indicated by their accumulation of ammonia and amides. Nevertheless, the actual fact of *de novo* synthesis of enzyme protein is hardly subject to doubt. This applies mainly, however, to the plant tissues adjacent to the point of infection.

organism are carried to the point of infection. Vavilov [1935] wrote of this, emphasizing the inadvisability of direct analogies between animal and plant in the matter of antibody production.

The absence of antibodies still does not indicate that in plants generally no acquired or induced immunity can exist. Such a possibility is admitted by many investigators [Ray, 1901; Beauverie, 1901; Vavilov, 1935; Müller and Börger, 1940; Cruickshank and Manruk, 1960; Lovrekovich and Farkas, 1965]. But induced immunity has much less importance in plants than in animals and is regulated by different mechanisms.

In the work of Weber and Stahmann [1966] it was shown that infection of a susceptible sweet potato variety with a nonpathogenic isolate of *Ceratostomella fimbriata* brought about immunity to a pathogenic isolate of the same parasite and also to a number of other potato parasites. The observation led the authors to conclude that resistance of sweet potato to several pathogens is based on certain common defense reactions.

In conclusion, it should be emphasized that, despite the great importance of antibiotic substances to the phenomenon of phytoimmunity, this role may not always lead by any means to a direct dependence between their quantitative content in plant tissue, enzyme activity responsible for their decomposition, and degree of resistance to the parasite. From the examples cited, it follows that this dependence is extremely complex and is observed most often after the interaction of plant and parasite.

Thus, the problem of immunity may be solved on the basis of a detailed study both of the nature of the chemical substances that play a protective role against infection and the biochemical processes responsible for their formation and conversion. Underestimation either of the role of a substance as an agent of two biological systems (parasite—plant) or of the biochemical processes determining their interaction can detract from an understanding of the nature of phytoimmunity and from the solution of a number of practical problems in the protection of crops from infectious disease.

A fundamental understanding of the biochemical processes occurring in plants in response to infection will frequently lead to the development of new protective substances or to the synthesis

of compounds already present in intact tissue that possess antibiotic properties against an invading parasite. Without knowing
which substances lie at the basis of the defense reaction of plants
and the means of attack on the parasite, it is impossible to begin a
study of the mechanism of interaction between them on a molecular
level, the importance of which in biochemical investigations is
generally admitted.

If definite progress in the study of plant antibiotic substances
has been made in recent years, there are so far not even generally
accepted approaches to a study of the toxic products of parasites
by which they accomplish their destructive action. Material concerning this question will be considered later, using as concrete
example the pathogen of tracheomycotic diseases.

Chapter II

Phytoncides and Phytoalexins

Phytoncides, according to Tokin's definition [1966], are
"bactericidal, fungicidal, and protistocidal substances produced by
plants as one of the factors of immunity with a role in the interactions of organisms in biocenoses" (p. 20). Phytoncide activity of
plants is demonstrated by many long-known facts concerning the
mutual effect of plants on one another — depression of life activity
in some and stimulation in others. It has been shown experimentally that volatile phytoncides of wheatgrass and oats stimulate the
germination of pollen grains of lucerne and, on the other hand,
volatile phytoncides of timothy exert an inhibiting action.

At first it was proposed that phytoncides are volatile substances formed by plants upon traumatization. Later, however, it became evident that nontraumatized plants also form phytoncides
which may be both volatile and nonvolatile compounds. The discovery of phytoncides belongs to the great achievements of modern
biology. But now there is no doubt that all plants, without exception, possess phytoncide activity. Contradictory results in the
literature concerning phytoncide activity of the same plants are
explained by the fact that it depends on numerous changes in the
physiological state of plant tissues that are not always taken into
consideration. For example, phytoncide activity of pine is twenty
times higher in summer than in fall [Kovalenok, 1944]. Infection
of a plant with chlorosis may lower the phytoncide activity ten to
fifteen times [Dzhaparidze and Kanchaveli, 1948].

The role of phytoncides in the phenomena of plant immunity
has been studied by D. D. Verderevskii and his colleagues for
many years. On the basis of their own studies and of a literature
review, Verderevskii [1959] reached the conclusion that, in the defense of the living body of plants against both parasitic disease
producers and a complex of saprophytic decay microbes, their
phytoncide properties are of prime importance. However, the experimental proof of this very important conclusion is complicated
by serious technological difficulties, since the chemical nature of
phytoncides has still not been adequately studied. The results of
investigations devoted to the chemistry of phytoncides have been
summarized by Drobot'ko et al. [1958], who noted that phytoncides
include various alkaloids, essential oils, polyphenols, and other
compounds.

Verderevskii [1957] worked out and used a method for the
study of phytoncides. Determination of the volatile fractions of
phytoncides, comprising, in the descriptive phrase of Kozo-
Polyanskii [1946], the first line of defense of plants, was performed
by examination of the effect of freshly ground plant tissues or of
juice extracted from them on spore germination and growth of
microorganisms. The experiments were carried out in Petri
dishes or watch glasses on the bottom of which was placed the experimental plant material and on the cover a spore suspension of
a suitable microorganism. For a study of the phytoncide activity
of the nonvolatile fractions, various methods for diffusion of the
sap from the experimental plant into agar medium sowed with the
microbe were employed. A study of the joint action of the volatile
and nonvolatile fractions of phytoncides was carried out by the
germination of spores in drops of sap extracted from the plant and
placed in Petri dishes on the bottom of which was pulp from the
ground tissues.

By this method it was shown that the pH activity of infected
plant tissue is higher than that of intact tissue, where for resistant
plant varieties these differences proved to be much greater than
for susceptible ones [Verderevskii et al., 1964]. The experiments
were performed with many plants and various pathogens. Thus, in
resistant varieties of corn infected with smut and also of grape infected with mildew, phytoncide activity almost doubled, and this
level was maintained for a long period. In susceptible varieties

the initial rise in activity was replaced by sharp recession. Upon infection of tomatoes with tobacco mosaic virus, a correlation was seen between reproduction of the virus and the antiviral activity of the tissues.

The importance of the investigations consists primarily in the fact that they showed convincingly a close relationship between the resistance of plants and their phytoncide activity dependent on the presence of antibiotic substances.

A number of questions arise, however, especially as to the reasons for the increase in phytoncide activity of plants in response to infection: is it due to a rise in the phytoncides present in the tissues even before contact with the parasite or to the formation of new antibiotic substances appearing only in response to contact with the parasite? Answers to these questions await further study of the chemical nature of individual phytoncides characteristic for different species of plants.

Many works have been devoted to a study of antibiotic substances originating in plant tissues in response to infection and of the so-called phytoalexins.

In a relatively short time, several phytoalexins have been isolated and chemically identified. This gave Cruickshank [1963], one of the leading investigators of phytoalexins, grounds for the statement that at present there is actually no need to clarify the nature of plant resistance by the presence of some hypothetical and still unidentified substance. And, although such an evaluation of the role of phytoalexins in plant immunity, is, in our opinion, somewhat exaggerated, the significance of the investigations is difficult to overestimate.

The theory of phytoalexins emerged in 1940 on the basis of experiments on the infection of potato tubers with a virulent and an avirulent race of *Phytophthora infestans*. The authors of these studies, Müller and Börger [1940], formulated their conclusions as follows.

1. Phytoalexin can be characterized as a factor specific for a plant that is formed *de novo* or upon activation through contact of the plant cells with a parasite.

Phytoalexin develops in the course of the necrotic reaction and is the cause of death of the parasite.

2. Phytoalexins can originate only in living plant cells.

3. Phytoalexins are chemical substances which are prod-
    ucts of necrosis of the host-plant cells.

4. Phytoalexins are nonspecific with relation to fungi,
    but different fungi are susceptible to them in differ-
    ent degree.

5. Resistant and susceptible plants show a similar quali-
    tative reaction to infection; the difference between
    them is confined to rate of formation of the phyto-
    alexins.

6. The defense reaction is concentrated in tissues con-
    taining the parasite and in tissues directly adjacent
    to the point of infection.

7. The rate of phytoalexin formation characteristic of a
    given plant is determined by its genotype.

These seven conditions, comprising the basis of the contemporary theory of phytoalexins, were formulated when the existence
of these compounds was still only suspected, since not one of them
had been isolated and chemically identified. Twenty-five years
have elapsed since then, and the above statements have not become
obsolete, although in recent times many new data have been obtained concerning the isolation, identification, and characteristics
of the phytoalexins of many plants.

Up to the present, the occurrence of antifungal substances
has been observed upon infection of sweet potato, orchid, barley,
carrot, potato, turnip, bean, green pepper, soybean, rice, and other
plants. From them have been isolated and chemically identified:
ipomeamarone, pisatin, orchinol, isocoumarin, trifolirhizin, and
phaseollin.

The first to be isolated, in 1943, was ipomeamarone [Hiura,
1943], which appeared on infection of sweet potato with *Ceratostomella fimbriata*. It was identified as 2-methyl-2-(4-methyl-2-
oxypentyl)-5-(3-furyl)-tetrahydrofuran [Kubota and Matsurua,1953].

Ipomeamarone

Orchinol

Orchinol is formed on the interaction of *Rhizoctonia repens* with tubers of *Orchis militaris* and is 9,10-dihydro-2,4-dimethoxy-6-hydroxyphenanthrene [Schellenbaum, 1959].

Upon interaction of carrot roots with *Ceratocystis fimbriata*, which is not a carrot parasite, isocoumarin was isolated. This compound is 3-methyl-6-methoxy-8-hydroxy-3,4-dihydroxyiso-coumarin [Sondheimer, 1961; Condon and Kuč, 1960; Condon, Kuč, and Draudt, 1963].

Isocoumarin

Pisatin

At approximately the same time, pisatin was isolated [Cruickshank and Perrin, 1960, 1961], forming in response to in-oculation of the endocarp of pea pods with *Monilia fructicola* and proving to be 3-hydroxy-7-methoxy-4',5'-methylenedioxychrom-anocoumaran [Perrin and Bottomley, 1962].

Trifolirhizin

Phaseollin

From the roots of red clover the glycoside trifolirhizin was isolated [Hietala, 1960], the aglycone of which is surprisingly reminiscent in structure of pisatin. Most recently, from bean in

response to inoculation with *Monilia fructicola* still another compound was isolated, called phaseollin [Cruickshank and Perrin,
1963]. Similar to pisatin, it proved to be 7-hydroxy-3',4'-dimethyl-
chromenochromanocoumarin.

Pisatin and phaseollin were isolated by the method of drop
diffusates, which, in outline, consists of the following: From the
valves of pea or bean the seeds were removed under sterile conditions and in their place in the seed chamber were placed drops
of water containing a spore suspension of the parasite. In the
water drop inside the seed chamber, the spores began to germinate. In response, the tissue of the endocarp formed phytoalexins,
preventing further growth of the spores. After a time the water
drops were harvested, the spores centrifuged out, and the liquid
exudate tested for toxicity to different parasites. For this purpose,
fresh spores were introduced into the exudate and their growth observed. Such a technique made it possible to obtain phytoalexins
without contamination with the contents of the plant cells, since the
membrane of the cell wall served as a unique filter.

The formation of phytoalexins was induced in the same plant
by the most diverse fungi, among which were facultative and obligate parasites, pathogens specifically adapted to this plant, and
phytopathogenic fungi not infecting the given plant species. Thus,
the formation of ipomeamarone in sweet potato was stimulated not
only by *Ceratostomella fimbriata* but also by *Thielavia basicola*,
Helmicobasidium mompa, and *Fusarium* spp. [Suzuki, 1957].

Orchinol was formed in orchid tissues both in response to
inoculation with *Rhizoctonia repens* and to infection with *Didymella
exitialis*, *Fusarium solani*, *Ophiobolus graminis*, and *Orcheomyces
bircini* [Gäumann and Kern, 1959]. The same applies also to isocoumarin in carrot tissue, the formation of which is initiated by
Ceratocystis ulmi, *Helminthosporium carbonum*, *Fusarium oxysporum*, and *Thielaviopsis basicola* [Hampton, 1962].

Cruickshank and Perrin [1962] compiled whole tables of tests
for the ability of different fungi to induce the formation of pisatin,
among which were both pathogens and nonpathogens of pea. No
regularity in the quantities of pisatin formed by pea in response to
inoculation with pathogens and nonpathogens could be observed.
However, since germination of all spores occurred under standard

conditions which, of course, were not optimal for all of the tested subjects, the authors are inclined to compare their formation of phytoalexins with great caution.

Nevertheless, all fungi without exception must not be considered to bring about the formation of phytoalexins. For example, orchinol was not formed when common saprophytes and semiparasitic soil fungi were used for the infection of orchids [Gäumann and Kern, 1959], and such a typical parasite as *Stemphylium radi-* . *cium* did not initiate the formation of isocoumarin in carrot [Condon and Kuč, 1960].

As an index of parasite susceptibility to phytoalexins, the value ED_{50} is introduced, i.e., the effective dose or that concentration of phytoalexin that inhibits the mycelial growth of the parasite 50%. The size of this dose is calculated from a plot of mycelial growth of the fungus as a function of phytoalexin concentration. That concentration of phytoalexin that inhibits the mycelial growth by half is the value of its ED_{50}. Each parasite has a characteristic ED_{50} in relation to each phytoalexin.

A test of the effect of phytoalexins on mycelial growth of different fungi proved that the fungi show varied degrees of susceptibility. It was established that phytoalexin specific for a given plant inhibits growth in culture of pathogens of this plant much less than that of nonpathogens. Thus, for example, *Ascochyta pisi*, a parasite of pea, is suppressed to a marked degree by phaseollin but is relatively resistant to pisatin; *Colletotrichum lindemuthianum*, infecting bean, is susceptible to pisatin but resistant to phaseollin.

On inoculation of the endocarp of pea pods with the pathogen *Ascochyta pisi* and with the nonpathogen *Sclerotinia fructicola*, pisatin is formed in approximately the same quantities. Tests prove, however, that a pisatin concentration of $2.8 \cdot 10^{-4}$ inhibits growth of the nonpathogen 100%, whereas growth of the pathogen is inhibited only 25% by this same concentration [Cruickshank and Perrin, 1960].

In the work of Cruickshank [1961], all fungi tested were clearly divided into two classes: those susceptible to pisatin and those relatively unsusceptible. The first group contained 44 parasites of other plants but not of pea. The ED_{50} for 38 of them was less than 50 $\mu g/ml$ (the exception was *Septoria pisi*, whose ED_{50} ranged between 75 and 100 $\mu g/ml$).

In this same work, results are presented on the growth inhibition of each fungus by pisatin at a concentration of 100 $\mu g/ml$. At this concentration the growth of five of the six tested parasites of pea is inhibited less than 50%, while 37 nonpathogens of pea are inhibited almost 90%.

If the concentration of phytoalexin observed in diffusates in response to inoculation with a fungus is compared with the susceptibility of fungi to that phytoalexin, as determined *in vitro*, it appears that nonpathogens of pea initiate the formation of phytoalexins in concentrations that exceed the value of their ED_{50}, and the quantity of phytoalexin formed in response to inoculation with pathogens is considerably lower than the value of their ED_{50}. In other words, the amount of phytoalexin formed in diffusates in response to inoculation with pathogens is insufficient to inhibit their growth. This conclusion was discouraging, since it indicated that phytoalexins play a protective role only in the phenomena of non-specific phytoimmunity, i.e., with incompatible combinations of host and parasite. Thus, it appeared that the phytoalexins themselves are responsible for the phenomenon of nonspecific immunity, as concisely expressed by Chester [1933], who states that the majority of existing plants are resistant to the majority of existing fungi.

In this connection, Cruickshank and Perrin [1965] undertook a test to establish the correlation between the quantity of pisatin forming in diffusates of different varieties of pea and the degree of their resistance. However, no dependence could be established. Further investigations showed a clear correlation between the amount of phytoalexin and the degree of resistance of the variety if the pisatin was determined not in drop diffusates, as generally done, but in the tissue underlying the endocarp from which it diffuses into the infective drop.

When spores of a nonpathogen of pea are placed in the drop, the developing hyphae penetrate the endocarp only in exceptional cases and remain mainly in the drop. In this case, the fate of the parasite is determined by the concentration of pisatin in the infective drop itself.

It is another matter, however, if the drops containing a spore suspension of a pea pathogen are placed in the seed chambers of the pods. Infecting hyphae that form after germination of such spores rapidly penetrate the tissue of the pod endocarp. Thus, the factor determining resistance upon infection with a pathogen is the concentration of pisatin in the tissues of the pea endocarp.

Calculation of the pisatin in the endocarp tissue showed that relatively resistant varieties formed two to four times more than highly susceptible varieties.

On investigation of pisatin formed in response to inoculation of the same pea variety with strains of *Ascochyta pisi* differing in degree of virulence, it appeared that the plant forms pisatin in much greater quantities on infection with a nonvirulent than with a virulent strain of the same parasite.

The activity of phytoalexins is determined by the degree of inhibition of the mycelial growth of the fungus and not of its spore germination. The fact is, the formation of phytoalexins begins only some time after spore germination in the infective drop, or, more precisely, in response to the metabolites that develop with germination. Thus, the effect of phytoalexins is directed not against spore germination but against the growth of the resulting hyphae.

As might be expected, hyphal growth proves much more sensitive to phytoalexins than does spore germination. Thus, growth of the mycelium of *Monilia fructicola* is three times more susceptible to the action of pisatin than is the germination of its spores [Cruickshank, 1965].

The formation of phytoalexins can be induced not only by the germinating spores themselves, but also by an extract after their growth. Thus, Uehara [1959] tested an extract after growth of Fusarium on the ability to develop phytoalexins. It proved to induce the formation of phytoalexins in the pods of soybeans. In experiments of Cruickshank and Perrin, it appeared that exudate obtained after sedimentation of spores that were 80% germinated brought about the formation of pisatin in much greater concentrations than the same exudate where only 1% of the spores had germinated [Cruickshank and Perrin, 1962].

Thus, the penetration of the parasite into the cell is not at all necessary for the induction of phytoalexin. Simple physiological contact of the metabolic systems of host and parasite is quite sufficient.

Nothing is yet known concerning the direct inductor of phytoalexin formation. It can only be assumed that the inductor is a substance of the nature of an enzyme or hormone.

Phytoalexins are formed also by a whole series of chemical compounds. Here, in first place, are ions of the heavy metals and also certain metabolic poisons. As early as 1960 it was reported [Uritani, Uritani, and Yamada, 1960] that ipomeamarone is formed in sweet potato tubers in response to treatment with mercuric chloride (0.1%), trichloracetic acid (5.0%), monoiodoacetic acid (0.3%), and 2,4-dinitrophenol (0.5%). In 1963, Condon, Kuč, and Draudt [1963] observed isocoumarin upon treatment of carrot tissues with certain chemical substances. Solutions of a series of chemicals induce the formation of pisatin in the endocarp [Perrin and Cruickshank, 1965]. It is interesting that low concentrations of chemicals do not bring about the formation of pisatin; with increase in concentration, pisatin development begins and increases, but with further increase in concentration, pisatin stops forming. Usually such high concentrations of chemicals possess marked phytotoxicity in themselves. In this connection, it is of interest to note that extremely high spore density in the infective drop also leads to marked reduction of pisatin formation [Cruickshank and Perrin, 1962].

Perrin and Cruickshank [1965] studied the ability to form pisatin of different cations associated with the same chlorine anion. It appeared that the highest ability to develop pisatin was shown by mercury, silver, and copper. Sodium, potassium, calcium, and magnesium were inactive. On the whole, the metals were placed in the following order of pisatin-forming ability:

$$Hg^{++} > Ag^{+} > Cu^{++} > Ni^{++} > Fe^{+++} > Cd^{++} > Zn^{++} > Co^{++} > Mn^{++}$$

A similar experiment was done with different anions combined with the same cation (K^{+}). According to their ability to form pisatin, the anions rank as follows:

$$ClO_4^{-} > NO_3^{-} > Cl^{-} > SO_4^{=}$$

It has been proposed that the inducing action of these compounds is explained by their ability to form complexes with enzymes containing sulfhydryl groups. This hypothesis is supported by the fact that many metabolic poisons, including monoiodoacetic acid and parachlormercurybenzoate (specific inhibitors of the sulfhydryl groups), induce the formation of pisatin. Apparently pisatin is not formed with normal functioning of the sulfhydryl enzymes in a healthy plant. The heavy metals, like metabolic poisons, are associated with the SH-groups of enzymes, which leads to an alteration in the isoflavonoid exchange (pisatin is chromanocoumarin, in the class of isoflavonoids), as a result of which pisatin formation begins. The hypothesis is supported by the marked reduction of pisatin initiation by parachlormercurybenzoate in combination with any sulfhydryl compound.

In our opinion, however, a number of facts are disregarded by this hypothesis. These include primarily the ability of a whole series of agents not directly affecting the activity of SH-enzymes to induce the formation of pisatin. Besides, if it is assumed that the destruction of activity of the sulfhydryl enzymes is the cause of pisatin formation, then it is not understandable why the compounds that themselves contain SH-groups, as, for example, thioglycolic acid and many others, induce pisatin formation. Apparently, interpretation of the mechanism of pisatin formation is still a matter for the future.

Many amino acids are effective with respect to pisatin formation, especially L-valine, DL-norleucine, and DL-norvaline. The D and L forms of the same amino acids exert different effects. L-valine has proved to be many times more effective than D-valine.

Kuč, Williams, and Shay [1957] and also Holowczak, Kuč, and Williams [1962] reported an increase in resistance of apple to *Venturia inaequalis* and of pea to *Aphanomyces eutriches* as a result of treatment with the same amino acids used in the experiments of Cruickshank. It is quite possible that the true cause of the increase in resistance of treated apple plants is the development in their tissues of phytoalexins responsible for the induced immunity.

Hence, a new possibility emerges for selection and testing of fungicides. Since most fungicides are compounds containing heavy metals, especially mercury and copper, their use in the treatment of plants should promote the formation of phytoalexins. Therefore, the effect of fungicides is twofold: on the one hand, they act directly on the disease agent and, on the other hand, they independently raise the resistance of a plant against the cause of infection [Cruickshank, 1965].

It is difficult at present to visualize the biogenesis of phytoalexins since those that have been identified belong to many different chemical classes. Apparently the formation of phytoalexin is the result of metabolic dysfunction of host-plant cells by a fungus infection.

The most detailed investigation has been on the question of biosynthesis of ipomeamarone-furanoterpene, the initial product in whose biosynthesis is mevalonic acid. It is proposed that in a normal plant CoA-acetate enters into the Krebs cycle, where it is oxidized to carbon dioxide and water. In the infected plant, acetate is converted to mevalonate and further, through peptinyl-, geranyl-, and farnesylpyrophosphate, to ipomeamarone. The final step in formation of ipomeamarone is the stage of enzymatic oxidation.

Isocoumarin is known to be synthesized from acetate on disruption of the normal metabolic flow by secretions of parasites [Condon et al., 1963].

As concerns the localization of pisatin in the plant, Cruickshank and Perrin [1965] were unable to observe a gradual decrease in its concentration with distance from the point of infection. Pisatin was not observed in the underlying mesocarp tissue, whereas in the endocarp tissue its concentration was 476 μg/g of tissue dry weight.

Quite another picture is characteristic of orchinol, which was observed throughout the infected tuber of orchids [Gäumann and Hohl, 1960].

Of great interest were experiments performed by Imaseki and Uritani [1964] on sweet potato infected with *Ceratostomella fimbriata*. Cylinders were cut from the tubers of the infected potato with a corer in a direction perpendicular to the inoculation surface. The first layers of the cylinder contained necrotized cells in which the fungus was observed. In the succeeding layers the parasite was absent.

The tissue cylinders so cut were divided into two parts. One part was cut into disks 0.5 mm thick; the second remained intact. Different layers of the cut cylinders from the first portion and also uncut cylinders from the second were incubated with radioactive acetate which served as source material for the synthesis of ipomeamarone. After incubation, the whole cylinders were also cut into disks. The radioactivity of the ipomeamarone extracted from each series of disks corresponding to the same layer was then measured.

It was found that, on incubation of individual disks cut before the start of the experiment, the greatest quantity of labeled ipomeamarone was in the fourth layer counting from the point of infection. The first layer in this case proved to be entirely incapable of including the label. On incubation of the entire cylinders, the labeled ipomeamarone was seen in largest quantity in the first necrotized layer of the tuber. The results justified the conclusion that ipomeamarone, or its precursor, is most actively synthesized in the external healthy portions of the tuber removed from the point of infection. Thus, ipomeamarone (or its precursor) migrates to the point of infection, where it accumulates in considerable quantities. The necrotic layer of cells thus appeared entirely unable to synthesize ipomeamarone when labeled acetate was used as source material. However, this could occur if some hypothetical substance X, the ipomeamarone precursor, is formed in the healthy portion of the tuber and converted to ipomeamarone in the directly infected tissue. It was demonstrated that the formation of ipomeamarone from precursor X in infected parts of the plant takes place much more rapidly than in tissues removed from the point of infection.

Thus, uninfected parts of the plant possess the ability to accomplish the step from acetate to the formation of precursor X and also the conversion of substance X to ipomeamarone. Directly infected tissue may form ipomeamarone only from precursor X, but it does this with much greater speed than uninfected tissue. Some sort of division of labor is effected between these two types of tissue. Uninfected tissue, using acetate as raw material, forms precursor X, which is then transported to the point of infection, where it is converted to ipomeamarone, which then suppresses growth of the parasite.

The formation of phytoalexins depends on the general physiological state of the plant. Strong young plants form phytoalexins more rapidly and in larger quantities than old, diseased ones. Mizukami [1953] compared the ability of the youngest (upper) and the oldest (lower) leaves of barley to form phytoalexins. It was found that the old leaves lost that ability more quickly than the young.

On storage of pea pods at 20°C, their ability to form phytoalexin was lost in six days, whereas at 4°C it was maintained at least 27 days. Anaerobic conditions also prevented the formation of phytoalexins. In all cases where the vitality of the plants is weakened, for example, on treatment with high temperature and narcotics, their ability to form phytoalexins is lost and susceptibility rises.

Phytoalexins are formed not only with artificial infection but also in naturally infected plants. The ability to form phytoalexins is peculiar not to any particular plant organ but to all its parts, to the plant as a whole. Thus, the presence of orchinol could be observed not only in tubers but also in the roots and leaves of orchids, and pisatin is seen not only in pea pods after their inoculation but also in infected leaves and roots.

From the above it is apparent that each plant, in response to inoculation with the most diverse pathogens, whether or not they are parasites of the given plant, as well as with many chemical substances, forms one or several phytoalexins peculiar to that plant. In this sense, phytoalexin is specific for the plant and does not depend on the nature of the parasite or chemical producing it. Depending on the nature of the inductor, phytoalexin may be formed in different quantity or at a different rate, but its nature remains the same. Apparently all plants within a species will form the same phytoalexin. Thus, in experiments of Cruickshank and Perrin [1965], it was found that all 58 experimental pea varieties formed pisatin. The difference between them was purely quantitative, i.e., only the concentration varied.

Moreover, it was shown that plants of closely related taxo-
nomic groups possess the ability to form the same phytoalexin in
response to infection. Thus, three species of pea (*Pisum sativum*,
Pisum arvense, and *Pisum elatins*) form pisatin. Orchinol is formed
by *Aceras antropophora*, *Anacamptis pyramidalis*, *Loroglossum
longibracteatum*, *Orchis morio*, and *Serapias neglecta* as well as by
Orchis militaris.

The following fact is of utmost interest. Of six observed and
identified phytoalexins, three have been found to belong to plants of the
legume family. These are pisatin from pea, phaseollin from bean, and
trifolirhizin from clover. Although all three phytoalexins are distinct
in structural details, they have as their basis the chromanocoumarin
skeleton. Apparently the relationship to one family, although it does
not determine complete identity of the phytoalexins formed, does to
some degree specify a community of structure.

Thus, phytoalexin structure is determined by the genotype of
the plant, not by the genotype of the parasite. The parasite genotype can determine only a certain degree of sensitivity to a given
phytoalexin.

A plant proves to be resistant only if it forms phytoalexin in
a concentration sufficient to inhibit the growth of the parasite and
susceptible if the concentration of the phytoalexin formed is insufficient to inhibition of the pathogen.

Phytoalexins suppress the growth of microorganisms at
average concentrations of the order of 100 μg/ml. If it is considered that antibiotics of the nature of penicillin and streptomycin
fully inhibit the growth of many streptococci and staphylococci at
very much lower concentrations, then the phytoalexins must belong
to the weak antibiotics. Such a conclusion was reached by Cruickshank [1961] for pisatin and by Gäumann, Nuesch, and Rimpau
[1960] for orchinol. The relatively weak action of pisatin, along
with some of its physical properties, limits our interest in its
therapeutic action. Its natural or synthetic analogs, however, could
represent a new source of biologically active substances capable
of a notable role in plant and possibly also animal chemotherapy.

Noting the great progress in the development of the theory
of phytoalexins, one must caution against an unwarranted attempt

to ascribe the resistance of plants to all pathogens, without exception, solely to the formation of phytoalexins. For example, not one
of the experimental bacteria in the studies of Cruickshank induced
formation of pisatin in pea in quantities capable of exerting an antibiotic effect. Consequently, some mechanism other than phytoalexin
must be sought to explain the resistance of peas to bacteriosis.

Plant resistance to phytopathogenic fungi also is by no means
always connected with the formation of phytoalexins. As noted even
by the authors of the phytoalexin theory, the role of phytoalexins is
manifested only in those cases where plant resistance is based on
the necrotic reaction. True, the necrotic reaction is widespread
in the plant world, but it alone does not determine the resistance of
plants to parasitic fungi.

Finally, even the necrotic reaction cannot lead to the formation of only a single phytoalexin. We have frequently noted that the
important role of antibiotic substances in phytoimmunity is due to
the presence in plants of not one, but several such substances, as
a result of which their action rises sharply. A whole series of results leads to the conclusion that not one but several phytoalexins
develop in infected plant tissue. Thus, in infective drops harvested
from the seed chambers of bean, the presence of at least two toxic
substances was observed, one of which was identified as phaseollin
while the other remains so far unidentified [Müller, 1958]. In addition to orchinol, Gäumann [1963, 1963/1964] observed another
toxic compound, hercynol, a derivative of orchinol, which like orchinol accumulates in the tissues of *Orchis militaris* but in notably
smaller quantities.

* * * * * * * * *

Thus, the investigators of phytoncides and phytoalexins join
in a single approach to the study of the nature of plant immunity
based on a recognition of the decisive role of plant antibiotic substances in the defense reactions of plants against pathogens of infectious diseases. And phytoncides and phytoalexins are considered to be plant tissue metabolites representing the most diverse
chemical compounds.

On this basis, the judgment is made that, in general, no difference exists between phytoncides and phytoalexins. It is well

known that the same phenomenon discovered in different laboratories, and certainly in different countries, often bears different designations for a long time, until one absorbs the other. We suggest, however, that this is not wholly a matter of terminology, although even the authors of the term "phytoalexin" recognize its imperfection.

The results of many experimental investigations lead to the conclusion that all antibiotic substances of plants can be divided into two basic groups from the point of view of their protective role in the phenomena of phytoimmunity.

The first group comprises antibiotic substances present in a plant before contact with a parasite. It is known that healthy tissues of many plants serve as raw material for the production of antibiotics. These substances are contained in plants in concentrations sufficient to suppress the growth of many microorganisms. They are products of normal metabolism of plant tissue and, therefore, can be called constitutional inhibitors. The content of these inhibitors can increase in the process of infection, as does that of other constitutional substances of the cell.

The second group comprises antibiotic substances that arise in plants after contact with a parasite. True, in individual cases some of these substances are seen in healthy plants but in such small quantities that they cannot exert antibiotic action. Only in response to infection do the substances of this group reach significant concentrations capable of suppressing fungus growth. Such substances, as a rule, are the result of altered exchange in which the host-plant metabolism shifts in response to the intervention of a parasite. Therefore, antibiotic substances of the second group may be called induced [Metlitskii and Akhvlediani, 1965].

Such a classification of antibiotic substances is provisional to some extent, since those contained in healthy and infected tissues are genetically related. But this applies as well to any other classification of chemical substances based on their biological action.

The above is not considered in the division of antibiotic substances into phytoncides and phytoalexins. As a rule, the investigators of phytoncides do not distinguish between antibiotics of in-

tact and those of infected tissue, but the investigators of phytoalexins consider only antibiotics produced in infected tissue and
disregard those contained in healthy plants.

In the interest of further profitable treatment of the problem
of plant immunity, it is imperative that a more precise terminology
be introduced and that workers in the area of phytoncides and phytoalexins join forces.

Chapter III

Energy Exchange and Phytoimmunity

Over fifty years ago Portier stated a hypothesis on the protective role of oxidases, according to which plant oxidases take part in the renewal of damaged integuments and also in the breakdown of toxins. "The Portier hypothesis," wrote A. N. Bakh in 1912 [Bakh, 1950], "which is accepted by other investigators, is very plausible and in all probability correctly explains pa rt of the functions of oxidative enzymes." (Bakh's emphasis.)

The role of oxidases in activation of the processes of restoration in epidermal tissues has subsequently been confirmed by many workers, and these results are now widely used in the protection of crops from infectious diseases (see Chapter IV). Much more rare are data on the role of plant oxidases in breakdown of the toxins of phytopathogenic microorganisms, about which we still know very little.

Upon infection of a great variety of plants with parasitic fungi, bacteria, viruses, and plant nematodes, it has been established with complete certainty that, in the great majority of cases, respiration rises sharply in infected tissues of the host plant. If the plant is successful in coping with the infection, respiration returns to normal. A drop in respiration is observed also with the development of disease, but in this case it is a result of the death of cells.

The universal identity in the character of reaction to the most diverse infections has long attracted the attention of investi-

gators. But precisely this identity of character, not only in response to interaction with different organisms but also to mechanical wounding and many other influences, calls for restraint in the
temptation to look for the mechanism of resistance to phytopathogenic microorganisms also at the level of peculiarities of plant
respiration, although such a correlation has been established in
many works.

In the comprehensive monograph of Rubin and Artsikhovskaya on the physiology and biochemistry of plant immunity [1960],
the results of their own extended investigations as well as numerous published data on the role of respiration in the phenomena of
phytoimmunity are reviewed. These studies furnish extremely
valuable information on the characteristics of respiration and the
activity of oxidative enzymes in resistant and susceptible varieties
of healthy and diseased plants.

The question, however, proved to be more complex than it
had at first appeared, since it is very difficult to distinguish,
among the many reactions arising in infected plants those that are
actually responsible for resistance and those that are artifacts.
For example, Kuč [1966] states that he knows not a single work in
which a change in respiration rate, respiration pathway, or enzymatic activity is directly responsible for resistance or can explain
its nature. Nevertheless, a detailed study of the role of respiration in the phenomena of phytoimmunity is of definite interest, since
respiration is not only an energy source but also the supplier of a
series of intermediate compounds for biosynthetic processes, including the formation of antibiotic substances. Such a study broad-
ens our ideas of the diversity of processes characterizing interaction of the two living systems of plant host and parasite.

A common index of the energy effectiveness of respiration is
the number of molecules of adenosine triphosphoric acid (ATP)
formed with absorption of one oxygen atom. This is expressed by
the coefficient P/O, the ratio of inorganic phosphorus absorbed to
absorbed oxygen.

Of course, oxidative phosphorylation does not proceed at the
maximum theoretical level in every cell at any given moment. The
P/O value expresses the energy requirement of the living cell and
may fluctuate in different stages of plant growth and development;
it also may depend on external conditions.

In living cells the rate of oxidative phosphorylation is controlled by the concentration in the tissues of adenosine diphosphoric acid (ADP), which takes on inorganic phosphorus, thereby
converting to ATP. The ADP concentration in the tissues is called
the respiratory control, since the rate of respiration depends on
it. If an acceleration of ATP utilization occurs in the cell, the
ADP thus freed removes the limitation on oxidation rate and respiration rises.

Respiratory control can be accomplished only if oxidation in
the cell is firmly linked with the phosphorylation process. In
some instances, however, this connection can become weakened,
and then part of the substrate is oxidized not by the phosphorylation pathway but by that of free oxidation. The energy of oxidation
in this event is not stored in macroergs of ATP but is dissipated
in the form of heat. This phenomenon by itself is not a guaranteed
index of pathology. It is used, for example, by animals in the process of thermoregulation [Skulachev, 1962].

The living system itself is able to regulate the degree of connection between its inherent oxidation and phosphorylation. Loss
or reduction of this capability leads to simple combustion of the
oxidation substrates, which occurs as though idling and is not accompanied by the formation of macroergs of ATP. Such respiration is known as "uncoupled" respiration, and substances bringing
it about are called "uncouplers."

Considerable evidence has been obtained for the uncoupling
of the processes of respiration and phosphorylation by reduced resistance of plant tissues to phytopathogenic microorganisms. As
an example can be cited our results in experiments on irradiation
of potato tubers and sugar beet roots with high doses of ionizing
radiation, weakening their resistance to parasitic fungi (Table 1).

Oxidation in irradiated tissues increases greatly, whereas
the phosphorylating ability rises less, thus leading to some lowering of the coefficient P/O. The oxidation substrate in irradiated
material operates with an appreciably lower coefficient of effective action, and part of the energy is liberated in the form of heat.

The external picture of respiration, usually judged by the
amount of absorbed oxygen and evolved carbon dioxide, does not
therefore represent the true character of respiration. It appears

Table 1. Effect of Gamma Radiation at a Dose of 50 krad on the
Process of Oxidative Phosphorylation in Mitochondria Isolated
from Potato Tubers and Sugar Beet Roots
(mitochondria from 20 g of dry tissue)

| Object | Variant | μatoms/hr | | P/O | P/O, |
		O_2	P		% of control
Potato tubers	Not irradiated	8.9	9.06	1.02	100
	7 days after irradiation	14.59	10.28	0.7	68
	Not irradiated	10.61	8.42	0.79	100
	90 days after irradia-tion	18.55	9.28	0.5	63
Sugar beet roots	Not irradiated	8.59	11.16	1.30	100
	7 days after irradiation	12.37	14.8	1.19	91

that the same external signs of respiration in infected tissue, including the activation of oxygen absorption, can equally indicate
such contrasting processes as the phenomenon of uncoupling, on
the one hand, and increased synthesis of macroergs, on the other.

With what kind of prerequisites is an increase in respiration
of damaged plant tissues associated? Compounds produced by the
parasite, or even products of metabolism of the plant itself, apparently serve as a direct stimulus to the increase in respiration. In
response, in the plant cells intensive processes of biosynthesis of
the enzyme systems, especially oxidative processes, are initiated.
New formation of enzymes is one of the reasons for increase in
oxidation and phosphorylation in a resistant variety. In addition,
the biosynthetic process of enzyme protein, like other synthetic
reactions occurring in this tissue, simultaneously determines another condition required for the rising oxidation: an adequate level
of ADP. The increased protein synthesis uses energy stored in
ATP, and this naturally raises the rate of ADP regeneration, lift-
ing the limit on respiration imposed by the ADP supply.

The increased respiration of the tissue of a resistant plant
is thus explained, on the one hand, by the new formation of enzyme
protein and, on the other, by the rate of ADP regeneration. A third
necessary condition is the presence of an oxidation substrate.
Starch serves as such a substrate in infected tissues of potato
tubers [Ozeretskovskaya and Metlitskii, 1966]. Therefore, in the
zone adjacent to a tuber wound starch grains disappear, with a

simultaneous increase in monosaccharides, which are also a sub-strate for the increased respiration.

Theoretically, three pathways for increase in energy output of living tissue can be postulated.

The first of these consists in a raised P/O coefficient, i.e., the formation of a large number of ATP molecules per unit of ab-sorbed oxygen.

The second pathway is based on an increase in the rate of electron transfer by the electron-transport chain (ETC) of mito-chondria, which leads to the development of a large number of ATP molecules per unit time. The rise in oxidation rate may be the result of sufficient ADP in the tissues to remove the limitation on the phosphorylation rate. On the other hand, this phenomenon may also be the result of removing from the electron-transport chain the phosphorylation step which had proceeded at the lowest rate and therefore inhibited the whole respiratory chain. In this case, a drop in the P/O value is inevitable, which may, however, be compensated for to a considerable extent by an increased rate of specific phosphorylation and formation of a large amount of ATP per unit time in a unit of mitochondrial protein.

Finally, the third pathway of increase in energy yield of the tissue is an increase in mitochondrial protein performing the process of oxidative phosphorylation, which correspondingly should raise the output of energy per unit of tissue weight.

It is evident that an increase in the energy output of damaged tissue need not be caused by only one of the above mechanisms, but may be the result of a combination of causes.

A study of the processes of oxidative phosphorylation in me-chanically wounded potato tubers [Ozeretskovskaya and Metlitskii, 1966] showed that in tissues of the zone adjacent to the wound the total protein content is much increased. A rise in the protein con-tent can be seen both in the composition of the supernatant fluid and in the mitochondrial fraction.

Phenol compounds interfere with determination of mito-chondrial protein in such subjects as potato by forming a dark-colored complex with the protein of the supernatant and possibly also with the mitochondrial particles themselves. On centrifuga-

Table 2. Protein Content in the Sediment of Mitochondria
Isolated by Different Methods from a Potato Tuber
(in mg/ 25 g dry tissue)

Variant	Additive	Protein content
Freshly cut tissue.	—	3.40
	PPO*	4.71
	Ascorbic acid	3.16
Tissue adjoining tuber wound† . . .	—	4.6
	PPO	5.53
	Ascorbic acid	4.0

*PPO = polyphenoloxidase.
† The zone adjoining the wound is specified as the layer of tissue 1 mm
  thick, cut from the wound surface after removal of the wound coating.

tion they sediment with the mitochondria, strongly affecting the results of protein determination. In order to avoid this error, we
tried to eliminate the phenols by combining them with a number of
adsorbents (caprone, perlone, silica gel, polyvinylpyrolidone). All
test compounds failed to bind the phenol compounds sufficiently.
In this connection, further isolation of mitochondria was always
carried out in the presence of the reducing agent ascorbic acid
(0.2 M), which eliminated the enzymatic oxidation of the phenols
contained in the supernatant (Table 2). The most protein was observed in the mitochondrial sediment isolated in the presence of
polyphenoloxidase (PPO). Evidently the oxidized phenol deriva-
tives formed by the action of PPO, along with denatured or cell
protein, sediment with the mitochondria. The purest mitochondrial precipitate was obtained in the presence of ascorbic acid in the
isolation medium.

According to our observations, the content of mitochondrial
protein in the adjoining wound layer rises markedly and remains
high for a week following wounding (Fig. 1A). The maximum content of mitochondrial protein was found in the adjoining wound zone.
Its quantity gradually falls with depth, and only at 7-8 mm from
the wound surface does it become normal (Fig. 1B). It is of interest that in both the first and second instance the curve of change

in mitochondrial protein is in full conformity with the respiration curve characteristic of this same tissue.

New formation of mitochondrial protein naturally poses the question of an increase in number of mitochondria. Increase in mitochondria in injured plants is noted in the works of many authors. In sweet potato tissues an increase in the number of mitochondria is mentioned by Uritani [1966] as a reaction to cutting the tuber. Lee and Chasson [1966] reported an increase in mitochondrial number in mechanically injured potato tubers. Similar results were obtained for tissues of tobacco infected with tobacco mosaic virus (TMV) [Weintraub et al., 1964; Rubin et al., 1966] and also for tissues of cabbage infected with *Botrytis cinerea* [Rubin et al., 1966]. In the latter case, the appearance of new mitochondria was observed in response to infection of both resistant and susceptible varieties.

It is very important to establish whether newly formed mitochondrial protein possesses the ability to regenerate energy at the same rate as the protein of mitochondria existing in the cell before injury. If this is the case, then the energy yield of such a cell should increase to the same degree as the mitochondrial protein.

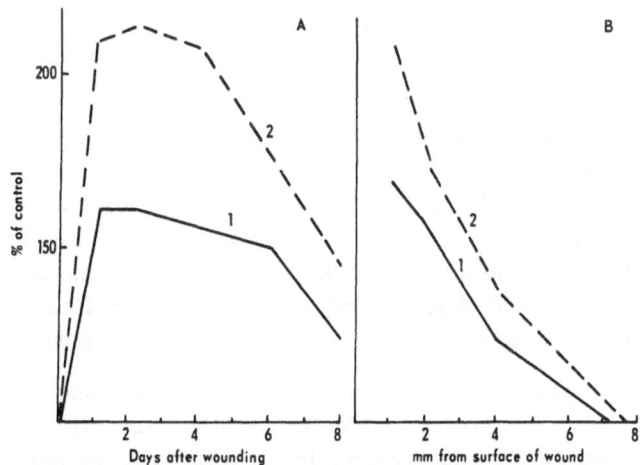

Fig. 1. Content of mitochondrial protein (1) and respiration (2) in a potato tuber on formation of wound periderm. A) In the adjoining wound zone of the tuber in proportion to wound periderm formation. B) At varying distances from point of wounding.

Table 3. Oxidative Phosphorylation of Mitochondria
Isolated from Freshly Cut Tissues of Potato Tuber
and from Tissues Adjoining the Wound
(mitochondria from 15 g of dry tissue)

Variant	Additive	μatoms/hr		P/O
		O_2	P	
Freshly cut tissue.	—	6.01	5.52	0.92
Tissue adjoining the wound	—	3.11	0	0
	Ascorbic acid	4.1	3.62	0.88

Our determinations indicated that the mitochondria isolated
from the adjoining wound layer without ascorbic acid possessed a
reduced oxidation ability, and their phosphorylation appeared completely suppressed (Table 3).

When ascorbic acid was added to the isolation medium, we
succeeded in obtaining phosphorylating mitochondria; however,
their activity lagged appreciably behind the controls. Nevertheless, the P/O coefficient appeared practically identical in the
freshly cut tissue and in that from the zone adjoining the wound.
We suggest that the reason for the low activity of the mitochondria
is the phenol compounds contained in the supernatant.

In this connection, later work was carried out with tissue
4 mm from the point of wounding. The reason for the use of this
particular tissue in the experiment was the fact that, according to
our findings, its phenol content is indistinguishable from that of
the control, while the mitochondrial protein still remains 25% ele-
vated. The presence of phenol compounds in these tissues, therefore, cannot distort the true picture of oxidative phosphorylation.

Table 4 presents our results on the oxidative phosphorylation of mitochondria isolated from the control (freshly cut) tissue
and from tissue of healed potato tuber 4 mm from the point of
wounding (experimental tissue).

The results obtained indicate that the supplementary mitochondrial protein is active in the formation of macroergs, and in
ability to regenerate them it is almost indistinguishable from the
mitochondrial protein of freshly cut tubers.

Calculations based on the observed data indicate that 15 g of dry tissue 4 mm from the point of injury forms 1.55 μM more ATP than the same amount of freshly cut control tissue. If it is assumed that mitochondrial protein in the zone adjoining a wound in the tuber has similar activity (its direct observation is precluded by the phenols present in the tissues of this zone), it can be concluded that the energy output in these tissues has risen more than 50%.

It should be noted that the P/O values in the control and experimental tissue 4 mm from the wound were almost identical. This justifies the assumption that energy necessary for the healing process materializes not by an increase in P/O but as a result of the additional formation of mitochondrial protein capable of active phosphorylation.

Akazawa [1956] observed that in sweet potato tissues adjacent to the point of infection with *Ceratostomella fimbriata*, the content of mitochondrial protein rises in comparison with that of the cut but uninfected tissue which in this case serves as control. The value of P/O in the tissue adjoining the infection point appeared notably higher than in the control. Along with a rise in the quantity of mitochondrial protein and an increase in the P/O value in infected tissues, the rate of oxidation and phosphorylation also rises. Evidently, electrons in infected tissue are transported more rapidly through the ETC and form a greater number of macroergs per unit time.

Table 4

	Control tissue	Experimental tissue
μatoms O/hr		
in 15 g dry weight.........	6.35	7.37
in 1 mg protein..........	3.97	3.68
μatoms P /hr		
in 15 g dry weight.........	6.04	7.59
in 1 mg protein..........	3.77	3.79
Ratio P/O...............	0.95	1.03
Content of mitochondrial protein (in mg)		
in 15 g dry tissue.........	1.6	2.0

A study of the processes of oxidative phosphorylation in the mitochondria of cabbage varieties resistant and susceptible to *Botrytis cinerea* [Rubin and Aksenova, 1964] showed that in the resistant variety, Amager, both oxidation and phosphorylation were much more strongly activated in response to infection and also the value of P/O rose, in comparison with the infected variety, No. 1. The authors came to the conclusion that the membrane and internal structure of the mitochondria were tougher in the resistant than in the susceptible variety.

The above emphasizes the extremely important function of respiration as connected with the formation and fixation of the energy of oxidation. Moreover, it would be incorrect to underestimate the other side of respiration, its participation in the process of plastic metabolism. It is known that, in the course of oxidation in the cell, a multiplicity of compounds arise which form the basic material for a different type of biosyntheses performed in the tissues.

Many of these compounds are unique, since the oxidation processes serve as the sole source of their formation. Among such compounds, for example, are erythroso-4-phosphate and pentoses forming in one of the steps of apotomic oxidation. The enzymes of this oxidation pathway, just like the other respiratory enzymes soluble in cytoplasm, take a most active part in the process of plastic metabolism.

As for oxidation in the mitochondria, the substrate entering into the mitochondria is fully consumed to the level of CO_2 and water [Ernster, 1957] and therefore cannot take any part in plastic metabolism. Of much greater importance in plastic metabolism is conversion of the substrate on the surface of the mitochondria by the pathway of free oxidation not connected with phosphorylation.

A rise in free as against phosphorylating oxidation naturally leads to some decrease in P/O. Such a reorganization, however, makes possible the attainment of plastic material necessary for the requirements of young growing tissues. Thus, Clerici et al. [1960] observed a rise in respiratory activity and a drop in P/O with regenerating liver. These experiments confirmed the view that the direction of metabolism toward biosyntheses is connected with activation of free oxidation.

As a result of the rise in free oxidation, part of the energy is produced in the form of heat. Increased heat production by the tissues of diseased plants has been observed by many: Eglits [1933] in potato infected with *Bacillus phytophthorus*, Fischer [1950] on infection of apple with *Botrytis cinerea*, and Yarwood [1953] in beans infected with rusts. It should be kept in mind that an increase in tissue temperature may also be a result of the uncoupling of respiration and phosphorylation which takes place on the development of necrosis, as will be discussed below.

The above-mentioned conversions in the oxidative system relate primarily to tissue of a resistant variety, where, in the process of oxidation, energy is formed and plastic substances develop in the course of defensive reactions of the plant.

As for tissues of susceptible plants, the general direction of the metabolic processes in this case evidently remains the same. However, the conversions cannot be realized, seemingly because of the action of parasite toxin which the tissues of a susceptible variety are not in a position to combat.

Entirely different is the role of energy exchange upon infection of plants with certain obligate parasites. Around the pustules of rust, for example, proteins, nucleic acids, organic phosphates, and many other compounds accumulate in great quantity and indicate a high level of energy exchange in the tissue [Gottlieb and Garner, 1946; Shaw and Samborski, 1956]. The same observations are made on infection of susceptible varieties of potato with wart caused by a typical obligate parasite, *Synchytrium endobioticum* [Lipsits, 1965]. A marked accumulation of organic products has also been observed at the points of injury in the tobacco plant by the tobacco mosaic virus [Shaw et al., 1954; Yarwood and Jacobson, 1955].

Among the metabolic products of susceptible plants infected with rust, a high content of growth substances was observed [Daly and Inman, 1958; Shaw and Hawkins, 1958]. These compounds may possibly be an indirect cause of the change in metabolism toward a predominance of synthetic processes. Such a stimulation of metabolic processes, however, is characteristic only for the early stages of infection. Later, a suppression of synthetic activity begins, which may be connected with the uncoupling of respiration and phosphorylation.

Perhaps the same mechanism by which a resistant variety realizes energy for the formation of protective substances lies at the basis of the initial intensification of energy exchange in susceptible plant varieties infected with obligate parasites. However, if in resistant varieties this mechanism provides resistance to the plant, on infection with obligate parasites the raised respiration provides for nourishment and energy of the parasite itself. Thus, according to the results of Lipsits [1965], the intense oxidative metabolism of potato infected with *Synchytrium endobioticum* does not serve the purpose of defense against the pathogen but, on the contrary, facilitates infection. This conclusion has been fully substantiated and is the regular rule, since a necessary condition for development and multiplication of the wart pathogen consists in the formation of meristematic tissues with active metabolism. Intensive oxidative metabolism provides the energy and metabolites required for the cell proliferation taking place in the infected tissues.

There is every ground to assume that the raised energy level of the tissues of susceptible plant varieties is a necessary condition for the successful development of obligate parasites, since many are themselves incapable of generating macroergs and are forced to use the energy of the host [Cutter, 1951].

The active energy system of the host plant that promotes its defense against many fungus pathogens can thus, in many cases, raise its infectibility for certain obligate parasites and viruses. The latter develop in tissues with active exchange, utilizing their energy and vital products. It is not by accident that plants resistant to fungus diseases in many cases prove to be easily infected with viruses.

A single characteristic of the energy exchange of a plant cannot, therefore, indicate the degree of its resistance, and certainly not to all parasites without exception; nor can it serve as a distinguishing feature for the selection of plants for resistance. It should be kept in mind that the significance of oxidative phosphorylation in infected tissue appears to be quite complex, and an approach to its determination should be made with great caution. Just as the general intensity of respiration based only on data of oxygen absorption and CO_2 evolution can give no idea of the real character of the actual processes going on, so also a one-sided determination of the P/O value does not reveal all the complexity of tissue energy exchange. In the living cell, in addition to an increase in the value of P/O to satisfy the need for energy, an increase in mitochondrial protein and a rise in rate of specific phosphorylation are still possible. On the other hand, some drop in the P/O ratio in a number of cases may indicate not an inhibition of energy effectiveness but enhanced processes of growth and biosynthesis.

Chapter IV

Plant Reactions to Wounding
and Their Protective Role against Infection

Long ago it was noted that in many plants, under favorable conditions of temperature, humidity, and aeration, a new tissue, wound periderm, forms at the point of wounding that is able to protect the plant from the penetration of an infection. A detailed study of the mechanism of this reaction is very important to an understanding of the nature of phytoimmunity as a whole and also to the development of methods for its intensification for the purpose of protecting crops from many infectious diseases.

With the harvesting of potatoes by combines, without which a large modern farm could not be operated, the infliction of mechanical injuries to the tubers is inevitable. For this reason, the principal source of loss in potato storage is the wound parasite *Fusarium solani,* a fungus which cannot enter through small cracks in the skin but requires more conspicuous wounds for penetration. Other microorganisms, which do possess the ability to enter the plant through undamaged epidermal tissue, nevertheless penetrate primarily through wounds. *Botrytis cinerea*, the main source of loss upon wounding of sugar beet, many species of vegetables, fruits, etc., may serve as an example. Intensification of the defense reactions in response to mechanical wounding would thus serve the purpose of protecting not only against wound parasites but also against other causes of infectious diseases.

Let us examine this question with potato tubers as an example, according to the results of our studies in cooperation with G. I. Chalenko.

The wound periderm of the tuber consists of several rows of elongated cells in a form reminiscent of brickwork, the covering of which is impregnated with suberin. Until recently, because of the lack of methods for appraisal of wound periderm and of degree of suberization, it was necessary to be satisfied with a mere calculation of the number of periderm layers and with a visual, purely qualitative reaction to suberization shown by the intensity of staining of the wound layer with gentian violet (GV). The results of a histochemical study of the wound reaction led to the conclusion that these two indices for the characterization of wound periderm are entirely inadequate.

For a quantitative determination of suberin, we developed and used a method essentially as follows. The tissue of the periderm wound layer* previously stained with gentian violet and repeatedly washed from excess stain is extracted three times with chloroform. As a result, all the stain that had combined with the suberin is transferred to the chloroform. The chloroform extracts obtained are read colorimetrically. The stain intensity is proportional to the amount of suberin present in the extracted coating. The results are converted to μg gentian violet combined with suberin from 100 mg dry weight of wound coating, according to a calibration curve for pure stain soluble in chloroform.

Only one determination of suberin quantity is insufficient, however, for a final conclusion as to the course of the wound reaction. From the microscope drawings presented (Fig. 2), it is seen that the formation of wound periderm proceeds unevenly in different parts of the tuber. It is formed most vigorously in the zone of the vascular bundles and is barely noticeable in the pith. A uniform wound periderm forms in the zone of the inner phloem region lying between the ring of vascular bundles and the medullary rays. It was this particular zone that we chose for experimental work. The fact is that the parenchyma of this part of the tuber is

_____

* By wound layer we understand the upper, predominantly corky layer of the periderm, which is easily separated from the surface of the healed tuber.

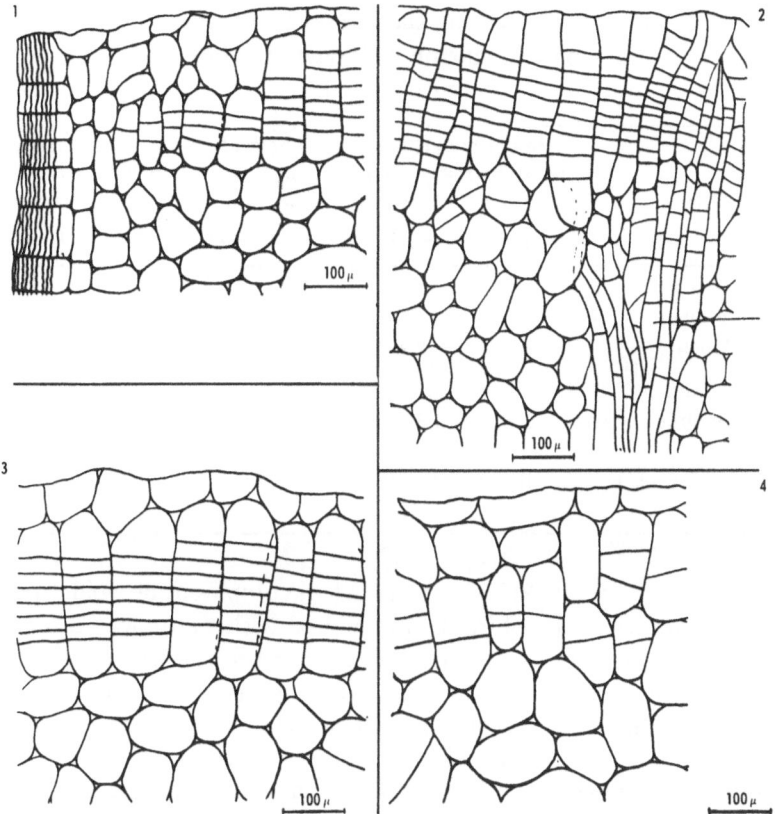

Fig. 2. Formation of wound periderm in different parts of a potato tuber. 1) Layer adjoining the natural skin (at left is seen the natural periderm of the tuber). 2) Zone of the neighboring vascular bundles. 3) Zone of inner phloem. 4) Zone of the pith.

relatively uniform in size and form, and we were convinced that the structure of the wound periderm developing from it depends on this to a considerable extent.

Sections prepared from tissue of the inner phloem of the tuber were stained with hematoxylin, which clearly distinguished the boundary between the suberized and the living part of the periderm. The stained sections were photographed and the negatives projected on paper, where measurements of the size of the peridermal cells were made. The results were then treated statistically.

Table 5. Effect of Washing the Surface of a Potato Tuber Slice*
on Formation of Wound Periderm
(with $r = 0.95$)

Variant	Height of vertical periderm rows (in μ)		Width of vertical periderm rows (in μ)	Number of layers	Suberin (in μg GV per 100 g of dry weight)
	Suberized portion	Living portion			
Unwashed sur-face of tuber slice (control)	97.0 ± 0.54	98.2 ± 1.48	84.3 ± 0.5	6.17 ± 0.2	35.2
Washed surface of tuber slice	106.4 ± 0.78	120.6 ± 1.68	106.6 ± 0.4	4.57 ± 0.04	23.5

*Surface of tuber repeatedly washed with tap water and periodically dried with filter paper.

Haberland [1922] observed that development of wound periderm in small slices of kohlrabi washed with water was slower than in similar, unwashed slices. This led him to the assumption that washing removes something from the slice surface that stimulates the formation of wound periderm; this he designated as wound hormones. A wound hormone was first isolated from young beans [Bonner and English, 1938] and called traumatin or traumatic acid. Traumatin was identified as decen-1,10-dicarbonic acid [English et al., 1939]. The presence of traumatin is characteristic for plant tissue; animal tissue does not develop it. Traumatin has been observed in potato, brussels sprouts, oranges, lemons, tomatoes, etc. Thus, in response to mechanical wounding, the same substance is formed in a great variety of plants.

Traumatin may quite possibly take part in the development of natural periderm, since the direct cause of its formation on the surface of a plant organ is a rupture in the covering cells of the epidermis associated with an increase in the dimensions of the organ.

The presence in the wound zone of substances stimulating the healing processes was shown by us in the course of the following experiment. The freshly cut surface of a potato tuber was treated with an extract from wound and adjacent tissues of a tuber. The largest amount of suberin was observed on the surface treated with extract from the wound layer. Treatment with extract from the adjoining zone did not affect suberin formation.

Our results indicated that washing the surface of a potato tuber slice with water markedly inhibits the process of suberization (Table 5).

Observations during wound periderm formation showed that the periderm developing on the washed surface of the tuber is characterized by several peculiarities. The number of its layers on the washed tuber surface is notably less than that on the control. Periderm that forms on the washed surface greatly exceeds the control in height of vertical rows, mainly because of the still nonsuberized portion. An outstanding property of such periderm is also the great width of its vertical rows as compared to the control.

The reason for the greater width of the cells of wound periderm under conditions unfavorable for the occurrence of the wound reaction (in the case under consideration, washing of the surface of the slice with water) may evidently be explained as follows. In the course of the wound reaction, each parenchyma cell lying under the wound initiates a row of peridermal cells. Under favorable conditions, all or almost all parenchyma cells undergo division. In this case, the width of the "brick" of the wound periderm is approximately equal to that of its mother parenchyma cell. Thus, according to our determinations, the width of the cell in the parenchyma zone of the inner phloem area reached an average of 85-90 μ. The periderm cells usually reached the same width under conditions favorable for the healing process.

With unfavorable conditions, not every cell of the underlying parenchyma layer divides. In this event, the developing periderm is compelled to expand in width to close the rows of cells over the whole wound surface. Thus, a great width of periderm cells indicates poor healing conditions, since some cells of the underlying parenchyma layer fail to receive the impetus for cell division.

The number of periderm cells was calculated as the arithmetic mean. Naturally, it does not imply uniformity of distribution over the whole wound surface. Distribution curves for the periderm layers can be constructed which show that the wound periderm on the washed tuber develops extremely unevenly (Fig. 3).

Alongside portions where multilayered periderm is found are rows of only one or two layers. A parasite will penetrate such areas much more easily, of course, than tissues with a stronger periderm barrier.

Our observations indicated that phenol compounds contained in the wound zone of a potato tuber, especially chlorogenic and caffeic acids, actively influence the healing process. Their effect (at a concentration of 1 mg/ml) proved to be opposite in character: caffeic acid inhibits the process of wound periderm formation, whereas chlorogenic acid stimulates it somewhat (Table 6). Caffeic acid removes to a considerable extent the impulse to division that is received by each cell of the parenchyma layer underlying the wound surface. Only isolated cells that are notably expanded in width undergo division. Treatment with chlorogenic acid does not affect the breadth of the periderm as compared with the control, i.e., in both variants all cells of the underlying parenchyma layer divide.

Treatment with caffeic acid results in increased height of the living layer of periderm where suberization is retarded. Chlorogenic acid, on the other hand, promotes suberization of the cell walls. The cells of the external portion of the periderm (the phellem), therefore, become suberized immediately after forma-

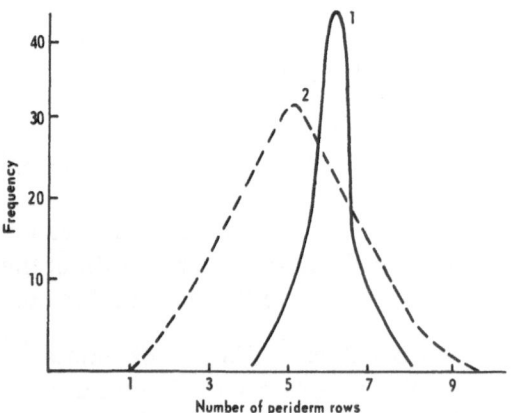

Fig. 3. Distribution curves of the number of rows of periderm on the surface of a potato tuber slice washed with water after wounding. 1) Unwashed surface. 2) Surface washed with water.

Table 6. Formation of Wound Periderm on the Surface of a Potato Tuber Slice Treated with Chlorogenic and Caffeic Acids (at a concentration of 1 mg/ml) (with $r = 0.95$)

Variant	Hgt. of vertical periderm rows (in μ)		Width of vertical periderm rows (in μ)	Number of layers	Amount of suberin (in μg GV) in	
	Suberized portion	Living portion			100 mg dry wt. of coating	100 cm^2 of the coating
Water (control)	111.54±6.22	137.8±3.64	99.1±0.52	5.7±0.2	57.5	102
Caffeic acid	129.6±9.0	164.6±7.6	122.3±0.32	5.2±0.6	33.6	69
Chlorogenic acid	108.2±6.7	83.85±4.12	99.3±0.5	5.9±1.8	75	122

tion. Under the microscope the wound periderm on the surface of the tuber treated with chlorogenic acid appears depressed and thickly packed. This is the cell form closest to the structure of natural periderm. Despite the relatively low height of the suberized layer of the tuber treated with chlorogenic acid, it contains the most suberin, whereas in periderm obtained after treatment with caffeic acid the suberin content is very low.

Microscopic examination of the potato tuber sections treated with caffeic acid showed that the wound periderm was embedded much more deeply than in the control. Thus, in 10 cases out of 100, in the control the periderm had been inserted in the very first row of the underlying parenchyma, and in the other 90 cases in the second row. In sections treated with caffeic acid, in no case was the formation of wound periderm observed in the first layer; it was most frequently in the third, and sometimes even in the fourth layer of the underlying parenchyma. In agreement with this, the depth of the periderm deposit in the control averaged 80 μ, whereas in sections to which caffeic acid had been applied it was 175 μ.*

* In this connection, the weight of the wound coating forming on the surface that was treated with caffeic acid is 1.5-2 times greater than the control. This should be taken into account in determining the suberin; its quantity in a given case is more correctly calculated per unit area of the wound surface (see Table 6).

Table 7

Radiation dose, rad	Amount of suberin
Control	55.1
10,000	44.0
50,000	32.6

The reasons for the inhibiting action of caffeic acid are of great theoretical interest. On the one hand, it is possible that caffeic acid binds (or prevents formation of) wound hormones responsible for cell division. On the other hand, it may be explained by the ability of caffeic acid to uncouple oxidation from phosphorylation, depriving the tissue of the energy necessary for the processes of cell division and suberin formation.

The latter explanation, participation of energy in the process of suberization, is still controversial. In the healing of a potato tuber irradiated with 10,000 rad, our results indicate that periderm formation barely occurs (only isolated cell divisions are observed), whereas the process of suberization is inhibited to a much lower extent. Data on the effect of irradiation on the suberization of the wound surface of a potato tuber (in μg GV/100 mg dry weight of the coating) are given in Table 7.

With a dose of 50,000 rad, which markedly lowers the value of P/O (see Table 1), suberization continues to proceed quite intensively. It would appear that suberization depends less on the energy supplied than does the process of wound periderm formation. However, if the formula for suberin composition is examined, it is difficult to imagine that such highly saturated compounds as the fatty acids which enter into its content could be synthesized without the participation of energy.

What then is the basis for the protective function of wound tissue? Is the wound periderm simply a barrier in the path of infection which retards the spread of a parasite only by means of its mechanical properties, or is this tissue also a unique chemical barrier containing compounds with antibiotic action?

A correct answer to this question has great significance for a better understanding of phytoimmunity as a whole. In a given case, the possibility is open to the investigator of studying plant tissue reaction, as it were, in pure form without the interference of a parasite and its inevitable secretions in the form of enzymes, toxins, auxins, and other physiologically active compounds.

Judging from the results obtained in our studies, the protective role of wound reactions cannot possibly be limited to the formation of a mechanical barrier. The developing barrier is, at the same time, chemical, since substances with antibiotic activity enter into its composition. This is affirmed by results of the following experiment [Metlitskii, Ozeretskovskaya, Chalenko, and Strokova, 1965].

A small piece of cellophane in the form of a cover glass was placed on a freshly cut slice of potato, on one following formation of wound periderm, on the same slice with wound layer removed, and, finally, simply on the wound layer removed from the tuber surface. On the cellophane was placed a drop of water containing a spore suspension of *Fusarium solani*. The potato slices, in moist Petri dishes, were then incubated at 15°C. After 19-20 hr, when the fungus spores had begun to germinate, the cellophane was carefully removed and transferred to a slide under the microscope, where the germinating spores were counted and the length of the infective hyphae measured.

With the presence in the tissue of low-molecular fungitoxic substances, the latter should penetrate the cellophane into the infective drop and suppress spore germination. The results repre-

Table 8

Variant	Number of germinating spores, % of control	Length of hyphae, % of control
Control.	100	100
Freshly cut tuber.	85	67
Healed tuber	63	46
Tuber with wound layer removed	61	44
Wound layer removed from the tuber	37	12

• All variants of the experiment had five replicates, in each of which 100 spores were measured. The same applies to all following tables containing results on calculation of germinating spores and hyphal length.

senting the germination of spores in the water drop on the cellophane placed on a slide are given in Table 8.

It appeared that in all experimental variants spore germination and hyphal growth were significantly suppressed as compared
with the control, the greater difference being in length of hyphae.
The strongest antibiotic action was observed in the last variant,
in which spore germination was reduced to almost one-third and
hyphal length to one-eighth compared with the control.

The results on the freshly cut tuber are of great interest.
Various compounds can penetrate through the cellophane into the
infective drop, not only those inhibiting but also those stimulating
parasite growth. Observations showed only the former, however.
Spore germination and especially hyphal length proved to be suppressed here also. In response to infection, evidently freshly cut
tuber tissue forms fungitoxic compounds of the nature of phytoalexins that inhibit spore germination and hyphal growth.

In the infective drop we observed a large quantity of phenols
which readily penetrate the cellophane because of the small size
of their molecules. Phenolic substances, as is well known, develop in large amounts in the wound and adjacent zones of a mechanically damaged potato tuber [Johnson and Schaal, 1957; Metlitskii and Mukhin, 1964].

Figure 4 shows the quantity of phenolic compounds in different parts of a mechanically injured potato tuber. The largest
amount was observed in tissues of the natural potato skin and the
next largest in the wound layer, where their content is many times
higher than in the parenchyma tissues of the tuber. The tissue of
the zone adjoining the wound is also rich in phenols, although here
their amount is considerably less than in the tissues of the natural
and wound skin.

The phenol extracts obtained were evaporated under vacuum
and the dry residue dissolved in water. In drops taken from such
solutions the *Fusarium solani* spores germinated. The length of
the developing infective hyphae proved to be inversely proportional
to the phenol content. The longest hyphae were those growing in
extracts from the parenchyma and the subcutaneous layer, the
shorter those growing in extracts from the natural and wound skin
(Fig. 4).

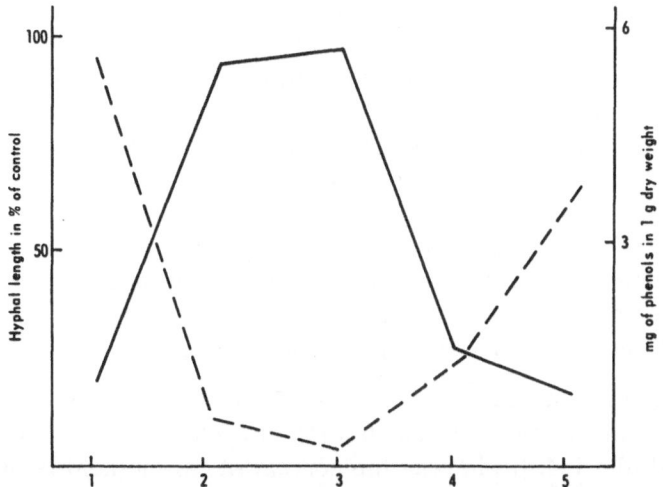

Fig. 4. Length of hyphae of *Fusarium solani* in extracts from different parts of a potato tuber (solid curve) and phenol content of these extracts (dashed curve). 1) Natural skin; 2) subcutaneous layer; 3) tuber paren- chyma; 4) layer adjoining wound; 5) wound layer.

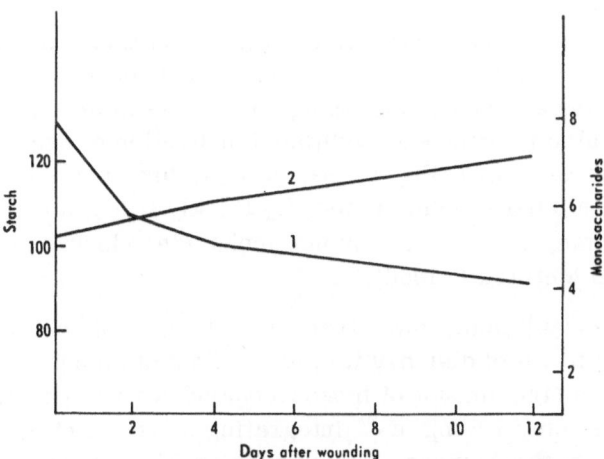

Fig. 5. Contents of starch (1) and monosaccharides (2) in disks of potato tuber in proportion to the formation of wound periderm (in mg glucose per gram initial dry weight of disk).

This does not, however, justify the conclusion that all fungitoxicity observed by us is due solely to the presence of phenolic
compounds. There is ground for suggesting that substances that
belong to other groups of compounds and also possess antibiotic
activity are formed in the wound and adjoining zones of mechanically injured tubers. Examples of such substances are solanin
and chakonin, steroid glycolalkaloids which accumulate in large
amounts in tissues of the wound zone [McKee, 1954, 1955]. A test
of their fungitoxicity [Ozeretskovskaya et al., 1968a] showed that,
at low concentrations of the order of 0.1-0.5 mg/ml, solanin and
chakonin inhibit the growth of *Fusarium solani* hyphae 70-90%. At
these same concentrations, caffeic acid exerts a considerably
smaller effect on the growth of parasitic hyphae.

Solanin is contained in tissues of the natural skin of the
potato tuber in an amount of 0.3-0.6 mg per gram of dry tissue
[Lampitt et al., 1943], whereas the content of caffeic acid is only
0.03 mg [Metlitskii and Mukhin, 1964]. If 1 g dry weight of potato
tissue is adjusted to 1 ml liquid, then it is apparent that in a physiological concentration of 0.3-0.6 mg/ml solanin strongly inhibits
growth and development of the parasite, whereas at a concentration of 0.03 mg/ml, like that of caffeic acid, it will not inhibit the
fungus development.

Mechanical injury to plant tissues is accompanied almost always by a rise in respiration [Kursanov, 1943; Prokoshev, 1943;
Ozeretskovskaya and Metlitskii, 1966]. Rising respiration of
wound tissue also presupposes additional utilization of oxidation
substrates. It was noted (Fig. 5) that in the wound zone starch,
with which potato parenchyma is usually thickly filled, disappears;
at the same time, the amount of monosaccharides rises [Ozeret-
skovskaya and Metlitskii, 1966].

Simple calculation showed that, after 12 days of healing, the
loss of starch in 1 g of disk dry weight consists of 38 mg of glucose.
In the same time the amount of monosaccharides rose only 1.9 mg.
In other words, only 1.9 mg of disintegrating starch is observed in
the form of monosaccharides. The remaining 36.1 mg apparently
was spent in respiration and also in the requirements of plastic
metabolism, i.e., in a different type of biosynthesis occurring in
these tissues. The results indicate that the preferred substrate
for the rising respiration of the healing tissue is the decomposing

Table 9. Effect of Potassium Cyanide on the Respiration of
Homogenates and on Oxidation of Succinic Acid
by Mitochondria of a Potato Tuber

Tissue	Conc., (moles)	O_2 absorption per hour			
		in 1 g dry weight of tissue homogenate		in mitochondria isolated from 15 g dry tissue	
		μl O_2	% of control	μl O_2	% of control
Freshly cut (control)	—	81.8	100	179.8	100
	10^{-3}	27.6	33.7	80.7	44.8
Adjacent to wound	—	53.4	100	120.5	100
	10^{-3}	40.7	76.2	72.5	60.2

starch. This, however, does not exclude the utilization of a whole
series of other compounds.

Injured tissue as a rule is characterized by qualitatively altered respiration. This is evidenced by results showing that the
respiration of mechanically damaged tissues becomes considerably less sensitive to many inhibitors that strongly suppress respiration in healthy plant tissues [Hackett et al., 1959; Calo et al.,
1957].

In our experiments it appeared that the respiration of the
zone adjacent to the wound in a potato tuber is considerably less
sensitive to potassium cyanide than is that of uninjured control
tissues (Table 9).

The respiration of tissue in the wound zone, according to
our determinations, also becomes much less sensitive to 2,4-
dinitrophenol (DNP) [Ozeretskovskaya and Metlitskii, 1966]. Such
an effect of DNP may be explained by the fact that the zone adjoining the wound in the tuber contains a considerable amount of ADP.
The latter is freed during biosynthetic processes of a different type
which proceed in these tissues with the aid of ATP energy. Such
processes are particularly those of protein regeneration, which
take place intensively in the course of tuber healing.

Protein was determined in healing disks cut from a potato.
From Fig. 6 it is seen that the protein content gradually rose in
proportion to the healing, and on the eighth day exceeded by one-
third that in a freshly cut disk.

In addition to processes of protein regeneration and synthesis of phenolic compounds in the wound and adjacent zones of the potato tuber, an accumulation of ribonucleic (RNA) and desoxyribonucleic (DNA) acids occurs [Metlitskii and Mukhin, 1964]. The largest amount of RNA is noted on the third day, after which it gradually decreases.

Biosynthesis of protein, nucleic acids, phenols, glycolalkaloids, and suberin, as well as processes of cell division connected with the development of wound periderm, require an additional supply of energy. We have shown that the needed energy is obtained through newly formed mitochondria in the zone adjoining the wound (see Chapter III for details).

Many substances that accumulate near the point of wounding prove to be physiologically active both for tissues of the plant which contain them and for hyphae of parasitic fungi penetrating the damaged areas. For example, caffeic acid, which suppresses the course of the wound reaction, also possesses fungitoxic activity. On the other hand, two substances, chlorogenic and caffeic acids, which accumulate in tissue in response to wounding, have opposite effects on the process of wound periderm formation.

Just how are the effects of these substances in the potato tuber reconciled? According to observations carried out by Metlitskii and Mukhin [1964], entirely different dynamics for chloro-

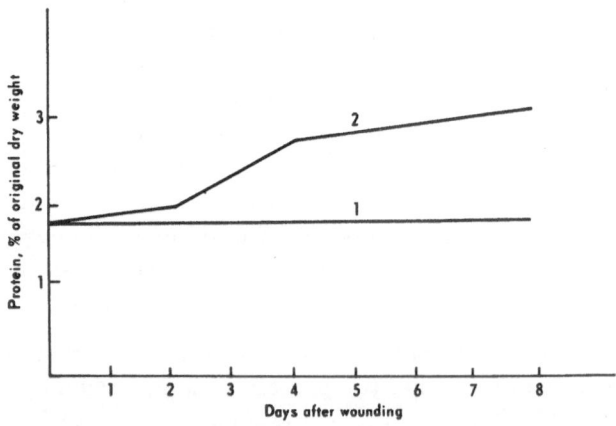

Fig. 6. Protein content in disks from a potato tuber in proportion to the development of wound periderm. 1) Freshly cut disks; 2) healing disks.

Table 10. Content of Chlorogenic and Caffeic Acids in a Potato
Tuber in the Course of Wound Periderm Formation
(in mg% of dry weight)

Number of days after wounding	Chlorogenic acid		Caffeic acid	
	wound zone	adjoining zone	wound zone	adjoining zone
3	19.6	30.3	3.2	0
7	16.2	50.1	4.3	0
14	13.6	30.3	4.3	3.6

genic and caffeic acids are characteristic for a wounded potato
tuber (Table 10).

Chlorogenic acid accumulates in the greatest amount in the
zone of the tuber adjoining the wound, i.e., just where cell division
is occurring. Caffeic acid, on the other hand, is concentrated in
the practically dead wound zone, where cell division is not taking
place. After seven days, when the wound periderm is essentially
formed, the content of chlorogenic acid decreases. As for the caffeic acid, while the processes of new tissue formation are proceeding, it is lacking in the zone adjacent to the wound. It appears
here later, when the periderm-forming process slows down, but
its largest quantity is observed in the wound layer, where with
other compounds it creates a fungitoxic barrier.

A study of the biochemical processes basic to defense reactions to mechanical wounding made possible the development of
methods for their intensification in storage of potato, sugar beet,
and some species of vegetables, for the purpose of protecting them
from infectious diseases and primarily from wound parasites. It
is a matter of employing, on a new biochemical basis, a known
method of storage under conditions of active ventilation formerly
applied advantageously for rapid cooling of the product to eliminate
self-heating. But for the intensification of wound reactions, a
completely different air exchange is needed and also another re-
gime of air temperature and humidity [Metlitskii and Volkind, 1966].

Chapter V

The Necrotic Reactions and the Role of the Polyphenol—Polyphenoloxidase System in Phytoimmunity

The necrotic reactions, also called abortive or hyperenergetic, or more often known as hypersensitivity, were first studied more than fifty years ago in the phenomena of wheat resistance to rust fungi. It was shown that the germ tubes of many rust fungi, on penetrating plants of immune varieties, very quickly bring about the death of the host cells with which they come into direct contact. It was noted that after a time certain substances diffuse from the host cells, killing the fungus hyphae. The higher the resistance of the variety, the smaller the zone of infection. In immune varieties, such a small number of cells die that the penetration of the pathogen can be observed only under the microscope. As a result, necrotic spots are formed whose size depends on the duration of the fungus growth and the number of killed host cells [Stakman and Harrar, 1957].

A similar succession of processes was established upon infection of different potato varieties that differed in their degree of resistance to phytophthorosis. According to the investigations of Tomiyama [1955, 1956, 1957], a cell of a Phytophthora-resistant variety dies during the first hour after penetration by the infective hyphae of the parasite. But the fungus hypha in the cell dies only after forty hours.

Two facts must be immediately stressed. First, by necrosis in these reactions is meant not death in general of cells and tissues upon development of an infection, but only "local death of cells under the influence of toxic products of decomposition" [Gäumann, 1954]. Second, necrosis is only a visible result of the defense reaction, and the development itself does not reveal the reason for the death of the plant cells and for the destruction of the penetrating parasite.

It has been suggested that necrotic reactions protect the plant only from biotrophic parasites, since such parasites are able to feed only on the contents of living cells and will starve in dead tissues. From this point of view, for facultative parasites nourished first by killed cells, the necrotized area should offer an even better substrate than living cells, whose resistance must first be overcome. However, experimental results indicate otherwise. The resistance of plants to many facultative parasites, and especially to *Botrytis cinerea*, has been shown to be connected also with the necrotic reactions. Destruction of the parasite occurs here not because of hunger but as a result of poisoning by the developing toxic compounds.

For a knowledge of the defense mechanism of the necrotic reactions, a study of the chemical nature and biological activity of the substances that emerge in the course of these reactions is extremely valuable. Does the death of the plant cells and of the penetrating parasite produce the same substances, or are they different?

Varied opinions on this question are presented in the literature. Some authors attribute the decisive role in necrotic reactions to phytoalexins, among which are substances of differing chemical nature. Others consider the basic mechanism of these reactions to be the polyphenol—polyphenoloxidase system present in the plants, by whose action oxidation products of the polyphenols accumulate in infected tissue and exert a toxic action both on the host cells and on the parasite.

Akazawa and Uritani [1961] investigated the change in degree of resistance of the sweet potato tuber to the black rot pathogen within a temperature range for the surrounding air of 20-34°C, and simultaneously traced the dynamics of formation of polyphenols and of the phytoalexin ipomeamarone. It was noted that the resistance of sweet potato correlated with the development of ipomeamarone but did not depend on the intensity of development of phenolic compounds.

Since phytoalexins have already been discussed, we shall now examine data obtained in studies of the polyphenol–polyphenoloxidase system.

It has been frequently noted that the content of polyphenols and the activity of polyphenoloxidase increase in response to infection; the greater the rise, the higher the resistance of plants to a pathogen. The question naturally arises as to the sources of the increase in polyphenol content; does it result from regeneration or from migration from adjacent healthy tissues? In our studies [Ozeretskovskaya and Vasyukova, 1965], a strong rise in polyphenol content was observed in a potato tuber disk in proportion to the development of wound periderm on its surface. Since the polyphenol content was calculated for the entire disk, migration from other tissues was excluded.

An indirect indication of polyphenol regeneration in response to wounding is the increase in infected and mechanically injured tissues of the processes of apotomic oxidation, which is the only pathway for formation of the erythrosophosphate necessary for the biosynthesis of aromatic rings [Rubin and Ozeretskovskaya, 1962].

However, the literature, along with reports of more energetic formation of polyphenols and stronger activation of polyphenoloxidase upon infection of resistant plant varieties as compared to those less resistant, also contains contradictory results. For instance, Lipsits [1965] noted that the activity of polyphenoloxidase and peroxidase was higher in infected tubers of potato varieties susceptible to the wart pathogen than in resistant varieties grown in an infected environment, where, on the other hand, the oxidase activity dropped. On infection of potato with the wart pathogen, polyphenol accumulation was seen only in tissues of susceptible varieties, where it was most pronounced in the wart tissue itself. On the basis of his studies, Lipsits came to the conclusion that polyphenolic substances, in the form in which they are concentrated in wart-infected tissues of susceptible varieties, are not merely nontoxic for the wart pathogen but, rather, promote its growth and spread through the tissues.

Table 11. Effect of Different Concentrations of
Chlorogenic and Caffeic Acids on the Hyphal
Growth of *Fusarium solani*
(in % of hyphal growth in a drop of water)

Concentration, mg/ml*	Chlorogenic acid	Caffeic acid
2	10	0
1	20	0
0.5	70	22
0.1	101	103
0.01	120	124

*The pH of the drop in which the tests were performed was 6.5
in all variants.

It must be remembered, however, that wart diseases are
actually distinct from all other diseases. And one of the peculiar
conditions emerging in wart tissue consists precisely in that, despite the increase in oxidase activity, the process of irreversible
oxidation of polyphenols is lacking. Since the wart diseases described by Lipsits are not accompanied by the formation of ne-
croses and, consequently, by damage to cell vacuoles and mixing
of their polyphenols with the cytoplasm, conditions for the opera-
tion of the polyphenol—polyphenoloxidase system are also lacking.

The majority of naturally occurring polyphenols in plants
possess only weak fungitoxic activity. An example is the fungitoxicity of chlorogenic and caffeic acids, which occur in a healthy
potato tuber. Table 11 gives our results in a test of the effect of
different concentrations of these acids on the hyphal growth of
Fusarium solani.

From these data it is seen that the inhibiting action of caffeic acid is somewhat higher than that of chlorogenic acid, but on
the whole the fungitoxicity of both is relatively low. At such low
concentrations as 10 µg/ml, both acids slightly stimulate growth
of the parasite hyphae, which is not unexpected since a similar action depending on concentration is a general property of antibiotic
substances.

Our observations [Ozeretskovskaya et al., 1968a] indicate that
caffeic and chlorogenic acids are greatly exceeded in fungitoxic

action by other substances contained in potato, the glycoalkaloids solanin and chakonin (Table 12).

Scopoletin, the aglycone of scopoline, also observed in tissues of the potato tuber [Hughes and Swain, 1960], appeared to be a strong inhibitor of the growth of fungus hyphae but suppressed spore germination to a lesser extent.

On the basis of a toxicity test of phenols alone, however, it is still impossible to judge the protective role of the polyphenol—polyphenoloxidase system. But when the whole system is considered, then not only the polyphenols themselves are involved but also products of their enzymatic oxidation in response to infection.

Model experiments have shown that the oxidation products of polyphenols possess much higher fungitoxicity than the initial compounds. However, to accumulate these oxidation products, the coupling of the processes of polyphenol oxidation and reduction, which is characteristic of the healthy cell and which is regulated by the corresponding enzymes, must first be destroyed. That is what happens in the course of the necrotic reactions. The cause may be the rupture of the proteolipid membrane of the tonoplast surrounding the cell vacuole, as a result of which the phenolic substances move out into the cytoplasm, where they undergo rapid oxidation [Sal'kova and Platonova, 1967]. Oxidized products of the phenols irreversibly inhibit the activity of a number of enzymes, and particularly that of dehydrogenases. Thus, in the course of the necrotic reactions in potato, not only a rise in polyphenoloxid-

Table 12. Concentration (in mg/ml) of Substances
Reducing Spore Germination and Hyphal Growth
in *Fusarium solani* by 50% (ED_{50})

Substance	Length of hyphae	Spore germination
α-Solanin	<0.1	0.45
α-Chakonin	<0.1	0.5
Caffeic acid	0.3	2.0
Chlorogenic acid	0.75	1.75
Scopoletin*	<0.1	>2.0

*Obtained from wormwood (*Artemisia persica* Boff.).

Table 13. Effect of Phenols and Their Enzymatic
Oxidation Products on Spore Germination and
Hyphal Growth in *Fusarium solani*
(as % of control)

Substance	Spore germination	Hyphal length
Water (control).	100	100
Chlorogenic acid.	103	58
Chlorogenic acid + PPO*.	35	46
Caffeic acid.	62	35
Caffeic acid + PPO	86	47
Pyrocatechol	100	97
Pyrocatechol + PPO	63	15
Catechol.	87	108
Catechol + PPO.	70	44

*Oxidation continued for 36 hr with a preparation of polyphenol-
oxidase (PPO) isolated from a potato tuber.

ase activity was observed but also a suppression of dehydrogenases
that catalyze reduction of polyphenol oxidation products [Rubin and
Aksenova, 1957].

The increase in fungitoxicity of some polyphenols during their
enzymatic oxidation may be demonstrated by the results of one of
our experiments [Ozeretskovskaya and Metlitskii, 1966]. It appeared that the fungitoxicity of the majority of tested compounds
increases notably after enzymatic oxidation.Results of the action
of phenols and their enzymatic oxidation products on spore germination and hyphal growth in *Fusarium solani* are given in Table 13.

Other authors [Linderberg, 1949; Oku, 1960; Stahmann, 1964]
showed that fungus growth was strongly inhibited by products of
enzymatic oxidation of gallic acid, pyrocatechol, and chlorogenic
acid. The same compounds in the nonoxidized state did not have
an inhibiting effect.

The studies of Farkas and Ledingham [1959] showed that, in
proportion to the enzymatic oxidation of pyrocatechol, its toxic effect on the growth of rust uredospores alternately disappears and
reappears, which may be related to the formation, in different
stages of oxidation, of compounds differing in their fungitoxic activity.

A different oxidation rate of polyphenols and the varied composition of their oxidation products that form in response to introduction of a parasite into resistant and susceptible varieties is apparently one of the reasons why the polyphenol—polyphenoloxidase system plays a protective role in resistant varieties but the development of the infection is not arrested in susceptible varieties, although its progress may be slowed down after some time.

Szent-Györgyi and Cietorisz [1931] noted that fungitoxicity of phenol oxidation products involves their nonspecific tanning effect on protein, as a result of which the latter becomes inaccessible to the parasite.

Model experiments have frequently shown that products of the enzymatic oxidation of polyphenols inhibit exoenzymes of the parasites. In one such experiment performed in our laboratory [Sal'kova and Bekbulatova, 1965], the effect of different phenolic substances on the activity of polygalacturonase from the culture liquid of *Botrytis cinerea* was studied. The phenols suppressed polygalacturonase activity most severely at the very beginning of enzymatic oxidation. Thereafter, as oxidation of the phenols was completed and condensation compounds had formed, their inhibiting action became far less severe. The action of polyphenols on polygalacturonase dropped substantially with addition of sodium diethyldithiocarbamate (DEDTC), blocking the polyphenoloxidase contained in the culture liquid.

Table 14. Effect of Phenolic Compounds (at a concentration of 1 mg/ml) on Polygalacturonase Activity (as % of control)

Substance	Effect of phenols		
	in process of enzymatic oxidation	after enzymatic oxidation	in presence of DEDTC
Caffeic acid.	30-50	10-15	15
Chlorogenic acid.	20-40	5-15	5-15
Ferulic acid	20-45	10	10
Catechol.	30-25	5-10	10
Gallic acid	20-25	5-10	10
Pyrocatechol	40-60	5-10	20-25
Tannin	40-50	5-20	10-25
Phloroglucine	2-5	2	2-5
Resorcin	2-5	2	2-5

Table 15. Effect of Certain Phenols (at a concentration
of 0.65 mg/ml) and of the Products of Their Enzymatic
Oxidation on the Absorption of Inorganic Phosphorus by
Mitochondria of a Potato Tuber
(mitochondria from 20 g of fresh tissue)

Variant	μatoms absorbed phosphorus
Water (control).................	9.7
Chlorogenic acid................	9.27
Products of its oxidation...........	6.71
Caffeic acid....................	4.85
Products of its oxidation..........	6.76

Of nine tested phenolic compounds, only 2-resorcin and
phloroglucine had no effect whatever on polygalacturonase activity
(Table 14).

Polyphenols, and especially the products of their oxidation,
are uncouplers of the processes of oxidation and phosphorylation.
It may be, therefore, that the toxic effects of these substances in
the necrotic reactions are indicated by destruction of the energy
apparatus of both the parasite and the host plant, as a consequence
of which both lack the necessary energy for their life activities.

With this objective, we determined the absorption of inorganic phosphorus by mitochondria from tissues of potato tubers in the
presence of caffeic and chlorogenic acids. The potato mitochondria,
even after repeated washing, contain active polyphenoloxidase,
which quickly oxidizes the added phenols. In order to prevent their
oxidation, ascorbic acid was added to the reaction medium (Table 15).

Caffeic acid proved to be the strongest uncoupler; it suppressed the absorption of inorganic phosphorus by half. Chlorogenic acid, on the other hand, barely affected phosphorylation. The
oxidation products of both acids appreciably suppressed phosphorus
absorption. It may be that the inhibiting effect of the products of
enzymatic oxidation is explained by their denaturing action on mitochondrial protein.

Direct determinations of oxidative phosphorylation [Metlit-
skii, Ozeretskovskaya, and Chalenko, 1965] in the necrotized zone
of potato tubers, variety Lyubimets (genotype R_1), developing in

response to infection with *Phytophthora infestans* (race 0), showed that the oxidative ability of mitochondria isolated from this zone is strongly inhibited, whereas phosphorylation is entirely absent (Table 16).

In order to show that the cause of the uncoupling is the phenol compounds present in necrosis, the following was done. To normal phosphorylating mitochondria obtained from healthy tissue of a potato tuber, an extract of phenols contained in the necrosis was added. As a result, there was considerable inhibition of the absorption of inorganic phosphorus and activatior of oxidation processes, i.e., the typical picture for uncoupling. After a preparation of polyphenoloxidase, which oxidizes phenolic compounds, was introduced into the reaction medium, mitochondrial phosphorylation stopped completely and oxidation was partially suppressed. Naturally, a cell in which energy storage is not taking place, oxidation substrates are exhausted, and enzyme proteins are irreversibly denatured quickly dies.

On the basis of the above, the conclusion that it is the polyphenol—polyphenoloxidase system that is primarily responsible for necrosis may be justified. It must not be forgotten, however, that some phytoalexins also can uncouple respiration and phosphorylation. Ipomeamarone, for instance, contained in necrotized sweet potato tissue infected with *Ceratostomella fimbriata*, destroys the energetics of the parasite [Uritani et al., 1954]. It is quite possible that phytoalexins, to a great extent, suppress the exchange process in the parasite and, to a lesser extent, affect the plant. Such a view is held by Allen [1966], who considers that the phytoalexins are not sufficiently toxic for plant tissues to account for

Table 16. Oxidative Phosphorylation of the Mitochondria
of Potato Tubers
(mitochondria from 15 g of dry tuber tissue)

Tissue	μatoms/hr		P/O
	O_2	P	
Healthy	8.26	7.44	0.9
Necrotized...................	3.45	0	0
Healthy + phenols	11.0	4.22	0.38
Healthy + phenols + PPO............	4.05	0	0

Table 17. Fungitoxicity of Extracts from Necrotized
Tissue of a Potato Tuber
(as % of control)

Extracted substance	Spore germination	Hyphal length
Cold alcohol (ethyl).	66	52.1
Hot alcohol (ethyl)	41	23.0
Chloroform*	18.5	7.0
Ether (sulfur)	0	0

*Tissue extracted with a mixture of chloroform and methanol (2:1); the
chloroform was evaporated under vacuum and the residue redissolved
in hexane.

the necrotic effect. In confirmation are the results of Cruickshank
[1963], who noted that phytoalexins developed in peas in response
to chemical poisons, but necrosis was not observed.

In the work of Müller [1958], however, convincing evidence
is presented indicating that phaseollin forming in bean inoculated
with *Sclerotinia fructicola* is responsible for the plant necrotic reaction.

Even granting that the necrotic reaction may be entirely due
to the action of the polyphenol−polyphenoloxidase system, nevertheless, in our opinion, it is not justifiable to ascribe to this system a determining significance in protective reactions of plants
against a penetrating parasite, since we have already mentioned
that the formation of necrosis does not reveal the actual mechanism of these reactions. Although products emerge in the course
of enzymatic oxidation of polyphenols that greatly exceed the ori-
ginal compounds in antibiotic action, their fungitoxicity in general
is relatively low (at least among those studied) and appreciably
lower than that of the phytoalexins described. This leads inevitably
to the idea that, along with the polyphenol−polyphenoloxidase system, another mechanism is involved in necrotic reactions that is
more active in relation to inhibition of the parasite itself.

In a study carried out in collaboration with N. I. Vasyukova
[Ozeretskovskaya, Vasyukova, and Metlitskii, 1968b] on the degree
of fungitoxicity of extracts from necrotized tissue of a potato tuber
(R_1) that developed in response to infection by an incompatible race
of Phytophthora (race O), it was shown that the extracts obtained

with alcohol, and consequently containing polyphenols, were notably exceeded in fungitoxicity by chloroform extracts, from which polyphenols are absent. Still more fungitoxic was an ether extract, which, at the same concentrations, completely inhibited spore germination in *Fusarium solani* (see Table 17).

To test the degree of fungitoxicity of substances insoluble in water, we used the following method. The text extract (corresponding to 50 mg dry tissue weight) was distributed evenly on the surface of a cover glass and dried for an hour at room temperature. A drop of aqueous spore suspension of *Fusarium solani* was put on a slide, which was covered with the cover glass (inner side). The slide was incubated for 18 hr at 15°C. The number of germinating spores and length of the infective hyphae of the parasite were calculated as percent of the control. The control variant was a water drop containing spores of the parasite covered with a cover glass, the inner side of which received the same amount of pure solvent and was then dried.

In Table 18 are presented data obtained by us on the fungitoxicity of chloroform and ether extracts from different tissues of a potato tuber. It is seen that extracts even from healthy tissue have some fungitoxicity. Greater fungitoxicity is noted for tissues of the wound layer, and the highest for extracts from the necrotized tissues. Even at a distance of 1 mm from the zone of injury, the fungitoxicity drops notably (in layers adjacent to the wound and to the necrosis). This once more indicates strict localization of the antibiotic substances originating in response to injury.

Table 18. Fungitoxicity of Extracts from Different Tissues
of the Potato
(as % of control)

Tuber tissue	Extract			
	Chloroform		Ether	
	Spore germination	Hyphal length	Spore germination	Hyphal length
Healthy.	84	60	96	70
Wound layer.	78	32	67	24
Adjoining wound. . . .	81	67	98	56
Necrotized.	15	7	0	0
Adjoining necrotized .	53	19	61	19

Chromatographic separation of ether and chloroform extracts by the method of thin-layer chromatography on silica gel showed that in tissues of the wound layer, and to a still greater extent in necrotized tissue, new compounds appear that did not exist in the healthy tuber.

Recently, Tomiyama et al. [1967] isolated and identified an antibiotic substance from necrotic tissue of a potato tuber that formed in response to infection with an incompatible race of *Phytophthora infestans* and that proved to be bicyclic norsesquiterpenoid alcohol supposedly of the following structure:

The compound was called rishitin. Its average effective dose (ED_{50}) against various races of *Phytophthora infestans* was $2.1 \cdot 10^{-4}$ M, which corresponds to 40 $\mu g/ml$ and approximates other phytoalexins in strength of action.

For comparison, we note that on testing the fungitoxicity of caffeic acid introduced into the culture liquid of *Phytophthora infestans*, it appeared that even at a concentration of 2 mg/ml, which exceeds its maximum solubility in water, the growth of the parasite was suppressed only 40-50% [Sokolova, 1964]. On microscopic study of the effect on hyphal growth of *Fusarium solani*, its action appeared stronger. The ED_{50} in this case was 300 $\mu g/ml$ [Ozeretskovskaya, Vasyukova, and Davydova, 1968a].

On the basis of the above, we propose that the protective role of necrotic reactions in the resistance of plants to phytopathogenic microorganisms is dependent both on the products of polyphenol oxidation and on other antibiotic substances, especially phytoalexins.

The role of polyphenol oxidation products cannot possibly be confined to necrosis formation. They doubtless act toxically also on the parasite and can thus weaken its resistance to phytoalexins. A synergistic action between the two might also be expected. Finally, polyphenol oxidation products can simultaneously localize the seat of the infection, protect from a second infection, and also forestall penetration into adjoining healthy tissues of the toxic sub-stances, which remain at the point of infection both from the dead parasite and from the destroyed host cells.

Further experimental study of these various questions will provide a clearer idea of the mechanisms basic to the necrotic reactions that play an important role in the resistance of plants to many pathogens.

Chapter VI

Biochemistry of Plant Resistance to Tracheomycotic Diseases*

In the preceding chapters we considered the chemical nature of individual antibiotic substances and the biochemical mechanisms of certain defense reactions. But, as already noted, resistance to a definite disease is most often determined neither by one antibiotic substance nor by a single defense reaction. Various substances can arise in response to infection and, in the course of disease development, several protective reactions can be set off at once. We shall examine this question with infectious wilt, one of the most widespread and damaging of plant diseases, as an example. Cotton is particularly vulnerable to wilt, but many other plants, herbaceous, woody, and shrubs, are also subject to this disease. The disease is caused by pathogenic fungi, bacteria, and viruses; we shall concentrate, however, only on the wilt diseases caused by fungi.

Despite the numerous differences in the characteristics of both the fungi and the host plants, the symptoms of these diseases have nevertheless much in common, as a result of which they are all combined in one group, the tracheomycotic diseases. Among the symptoms are injury to the water-conducting system, loss of turgor of tissues and their death, darkening and clogging of the xylem vessels, yellowing of the leaves, and, finally, shriveling of the plant. The community of symptoms justifies the assumption

---

* This chapter was written jointly with K. V. Vasil'eva.

of the presence of fundamentally identical defense reactions on the part of the plants, i.e., a uniform nature of wilt resistance. Knowledge of the general principles of resistance to tracheomycotic diseases should, however, facilitate a study of the characteristics of resistance to the individual pathogens of these diseases.

Many comprehensive reviews have been devoted to the physiology and biochemistry of wilt, the description of symptoms, and the explanation of its nature [Dimond, 1955; Gäumann, 1957; Ludwig, 1960; Sadasivan, 1961; Sukhorukov, 1963; Beckman, 1964].

The current hypotheses of defense reactions in plants with wilt are being formulated on the basis of reported experimental results and observations cited "for" and "against" two earlier theories explaining the main cause of wilt and the corresponding mechanism of resistance: the "plug theory" and the "toxin theory."

According to the first theory, the principal reason for wilting of a plant is the irreversible destruction of the function of the xylem vessels as a result of plugging with mycelium, polysaccharides, fungus secretions, gum and resin compounds, gels, and tyloses. The most widely circulated idea is that the active elements responsible for vessel plugging are the pectolytic enzymes of the parasite [Winstead and Walker, 1954a; Waggoner and Dimond, 1955; Pierson et al., 1955; Scheffer et al., 1956; Gothoskar et al., 1953, 1955; Kamal and Wood, 1956; Deese and Stahmann, 1962a, 1962b; Wood, 1959, 1961].

The pectolytic enzymes are suggested as the cause of breakdown of pectic substances of the plant cell walls, as a result of which large fragments of high-molecular polyuronides are freed. The latter, dropping into the transpiration stream, cause a rise in viscosity of the transpiration fluid that results in a plugging of the vessel lumen. The substances clogging the vessels may be calcium pectates which form upon combination of a calcium ion with pectic substances demethoxylated by the fungus enzymes.

It seems entirely probable that the pectolytic enzymes are also connected in some way with the formation of tyloses that fill the vessel cavities and cause their plugging. Tyloses develop by the protrusion of neighboring parenchyma cells into the vessel cavities through a membrane evidently weakened by the action of the fungus pectolytic enzymes.

Among the basic facts substantiating the theory of vessel plugging and the role of pectolytic enzymes are the following: formation of pectolytic enzymes by wilt fungi on cultivation *in vitro* ; formation of these enzymes upon cultivation of the fungus on a living stem and correlation of their activity in the tissues with degree of resistance of the plant to the disease; wilting of cut plants placed in preparations of pectolytic enzymes, either commercial or isolated from the culture filtrate of wilt pathogens; the same histological picture of vessel plugging in plants infected with wilt and in those treated with a culture filtrate of the fungus pathogen.

Very few results on the role of cellulolytic enzymes in the development of the disease are available [Talboys, 1958c; Husain and Dimond, 1960]. According to the opinion of Husain and Kelman [1959], the role attributed to pectolytic enzymes must be shifted to the cellulolytic, since the above-mentioned studies were carried out with preparations of pectolytic enzymes that contained cellulase.

A number of objections, however, are presented to the theory of vessel plugging, and consequently also against the role of pectolytic enzymes in disease development. First, the mechanical plugging observed in plants is not so important as to explain by itself the great reduction in the transpiration stream that takes place in diseased plants [Ludwig, 1952; Talboys, 1957]. Moreover, rapid destruction of the ion balance is observed in an infected plant, and this certainly cannot be a result of mechanical plugging of the vessels [Sadasivan and Kalyanasundaram, 1956]. Some results have indicated that preparations of purified pectic enzymes isolated from the culture filtrate of *Verticillium albo-atrum* did not cause plugging in the plant stem at all [McIntyre, 1965].

Mutants obtained by ultraviolet irradiation, which are unable to produce pectolytic enzymes, have been shown nevertheless to cause wilting of plants [McDonnell, 1958, 1962]. Pectolytic enzymes affected susceptible and resistant plants in the same way.

Wilting of plants can be caused by pectolytic enzymes contained in the culture filtrate of a fungus that is not among the wilt pathogens but causes rot of potato tubers [Fischer, 1965].

Such is some of the evidence that can be brought in defense or refutation of the plug theory.

According to the second theory, the principal cause of infectious wilt is toxins secreted by the pathogens in the xylem vessels that disturb the osmotic function in the cells of living tissues, especially in leaves. But, before citing evidence for this theory, it must be noted that in the literature of phytopathology different points of view regarding the very term toxin still appear [Kuprevich, 1947; Gäumann, 1957; Dimond and Waggoner, 1953; Ludwig, 1960; Akhvlediani, 1964; Braun and Pringle, 1959; Pringle and Scheffer, 1964; Wheeler and Luke, 1963; Allen, 1966].

In certain experimental work, the term "toxin" is used to include all culture liquid in which the parasite has grown or all metabolites, without exception, that are toxic to the plant and are formed by the parasite. But it is perfectly apparent that conditions of cultivation of parasites in artificial medium differ essentially from the nutritional conditions in a diseased plant. Therefore, in the culture liquid, along with true toxins, other substances may accumulate which will be damaging to the plant but which alone are not responsible for the development of disease. The parasite itself can form toxins strongly distinguished from each other in degree and character of action. Thus, according to the results of Akhvlediani [1966] on two toxins from malsecco, the pathogen of infectious shriveling of citrus, one (malsecco toxin A) caused plugging of the vascular bundles and wilting, whereas the other (malsecco toxin B) caused only local necrosis.

Because of serious technical difficulties connected with the isolation and study of toxins, our knowledge of their nature, structure, and biological action still remains rather fragmentary. A situation has developed where the absence of sufficient experimental data prevents a precise definition of the term "toxin" and the development of a classification of toxic metabolites; and, in turn, the absence of a precise determination and classification of toxins as to their role and position in the etiology of disease hampers experimental search for toxins and an evaluation of those already isolated. It is not by chance, therefore, that, of the large number of poisonous substances observed in media in which parasitic fungi have been cultivated and that have been described as toxins, only a few compounds have subsequently been assigned to true toxins, namely, those directly responsible for the appearance of the disease symptoms. But, with classification of isolated toxins and knowledge of their biosynthetic pathway, it will be easier to organize the search for and testing of substances able to prevent the formation of toxin and to render it harmless in the plant itself immediately upon formation.

Not without basis, Becker [1963] notes that fungus toxins have much in common with antibiotics both in the specificity of their action on definite metabolic chains and in the conditions of their formation. Their differences have to do with the fact that the effect of antibiotics is directed predominantly against microorganisms, whereas toxins affect macroorganisms, i.e., animals or higher plants.

Toxic metabolites have by no means just a single role and position in the development of disease. Toxins with selective action and causing all the symptoms of a disease are found only when rapid death of the plant or its parts is characteristic for the disease [Allen, 1966]. But often, from the initial effect of the toxin to the appearance of a visible characteristic symptom, a whole chain of reactions occurs involving metabolites of the plant host and others of the microorganism, as well as products of their interaction. It is, therefore, more expedient to look for toxins with selective action primarily in the early stages of the disease. In the later stages the symptoms are largely due to metabolites of the host and products formed by the interaction of parasite and host.

The results of studies on the role of toxins of the parasite in the development of infectious wilt are reviewed in a number of surveys [Dimond, 1955; Gäumann, 1951, 1954, 1957, 1958; Ludwig, 1960; Sadasivan, 1961; Subramanian and Saraswathi-Devi, 1959; Van den Ende, 1958; Beckman, 1964; Akhvlediani, 1964]. And, although some authors consider that the responsibility of each of the isolated toxic compounds for one or another symptom of the disease is already established, an analysis of the data on the properties and activity of the observed toxins leads to the conclusion that they could hardly be the only causal agents of the disease.

A search for toxins in culture filtrates of the wilt pathogen was undertaken on the basis of the hypothesis that wilting of cut plants is an adequate specific test and that metabolic products formed *in vitro* appear also in the diseased plant and take part in pathogenesis. The observed "wilt toxins," however, did not cause symptoms in the plant characteristic for wilt. They often recalled the disease only superficially. For example, the course of transpiration in a plant treated with lycomarasmin differs notably from that in a diseased plant. Diseased plants and those treated with lycomarasmin as well as with another toxin, fusaric acid, did not react similarly to treatment with iron solutions. Entirely different toxins may be responsible for the disease symptoms and not those that have been isolated and described up to the present time.

Gottlieb [1943], by centrifuging stem sections of a diseased tomato plant, isolated the transpiration fluid which caused wilting of cut tomato plants. The induced wilting was reversible; also reversible was the change produced in the semipermeability of the protoplasm. Data on the nature of the active principle, however, do not exist. But, according to the classification of Gäumann [1954], the outstanding characteristic of toxins consists in the irreversibility of their action. For example, lycomarasmin and fusaric acid irreversibly destroy the semipermeability of protoplasm.

No definite evidence could be obtained on the role in disease development of even such a well-known toxin as fusaric acid. Some results on the concentrations at which it accumulates in infected tissues indicate that it is not able to cause those physical and chemical changes that are observed in a diseased plant, for example, increased respiration, raised activity of polyphenoloxidase, and higher mitochondrial activity. Fusaric acid at these concentrations even produced the opposite effect: reduction of respiration, decrease in polyphenoloxidase activity, and lowering of mitochondrial activity [Wu and Scheffer, 1962; Kuo and Scheffer, 1964]. True, it is quite likely that these results indicate something entirely different. It is by no means impossible that the observed toxins exert an effect on the course of the disease by action on other processes and not on those that were studied.

At the present time, therefore, there are not sufficient grounds to consider toxins, at least those that have been observed, the sole cause of the disease symptoms. The lack of specificity in their action tends to support this statement: they exert the same effect on both resistant and susceptible plants. We must also note the absence of a correlation between ability to produce toxins and pathogenicity of strains. For example, the ability to produce

fusaric acid is shown also by members of the genus Fusarium that
do not cause wilt [Gäumann, 1958; Fischer, 1965]. It is still difficult to say anything about the presence in fungi causing wilt of specific toxins with a selective action. There are data showing
greater toxicity of extracts from the mycelium of some species of
Fusarium and Verticillium for the host plants than for plants they
do not parasitize, as well as data indicating the absence of such
selective action. For example, Winstead and Walker [1954b], by
subjecting intact plants of resistant and susceptible varieties of
tomato, radish, cabbage, and cotton to certain dilutions of culture
filtrates of the corresponding fungi from the genus Fusarium,
showed the possible existence of such selective toxins with low
molecular weight. From the culture filtrate of *Verticillium
dahliae* substances were isolated that exert greater toxic effect in
cotton varieties susceptible to this pathogen than in those that are
resistant [Krasil'nikov et al., 1965, 1966].

Davis [1965] showed the formation of certain metabolites, on
interaction of the wilt fungus pathogen with susceptible plants,
which by their action can make plants that are not hosts to the
given parasite susceptible to wilt. Hiroe and Nishimura [1956a,
1956b] propose the presence of a toxin with selective action in culture filtrate of the fungus causing watermelon wilt. But neither
phytonivein nor fusaric acid possesses such activity.

The question as to the presence of toxins with a characteristic selective action in the wilt pathogens is thus still unanswered.
And, certainly, on the whole, great care must be taken in the interpretation of results obtained in a study of the chemical nature
and biological action of toxins.

As a consequence of opinions expressed on pectolytic enzymes or toxins as the primary cause of wilt, corresponding pro-
posals appeared as to the defense reactions of plants.

Wilt resistance was connected with the ability of plant tissues to withstand the effect of the enzymes (or, correspondingly,
of the toxins) either by a special mechanism causing inactivation,
and possibly also inhibition, of enzyme production or by the de-
toxification of the toxins. According to the observations of a number of investigators [Talboys, 1957; Gäumann, 1951; Scheffer and
Walker, 1954], tissues of resistant varieties are subjected to the

action of the fungus metabolites just as are the tissues of susceptible varieties. In a resistant plant, however, the symptoms remain localized and it recovers from the injury.

From the above it follows that neither the plug theory nor
the toxin theory can explain either the nature of the disease or the
resistance of the plant. The results of numerous works, mainly
in recent years, lead to the conclusion that in the course of disease
development interactions take place between various metabolites
of fungus and plant, constituting different defense reactions.

Talboys [1957, 1958a, 1958b, 1964], on the basis of long
years of study of the wilt of hops caused by *Verticillium albo-
atrum*, reached the conclusion that the disease development has
two phases, determinative and expressive. In the first phase, it
is determined whether the fungus can penetrate the vascular system of the host and consequently whether it can maintain itself in
the plant. A decisive role is played here by interactions of the
penetrating fungus hyphae with the root tissues (epidermis, cortex,
endoderm). It was shown that the defense reactions which lead to
localization of the infection involve lignification of the cell walls
of the epidermis and cortex and also lignification and suberization
of the endoderm cells. Strong virulence of a parasite strain, according to Talboys, is associated not with the ability to overcome
this reaction but with the ability to inhibit its development. Similar
reactions were seen in cotton roots upon infection with *Verticillium
albo-atrum* [Presley et al., 1966].

The defense reactions observed in cotton roots include swelling of the cell walls and deposit of gummy and dark-colored substances [Garber and Houston, 1966]. With all these reactions originating in the roots upon introduction of the fungus, in the final
count the fate of the infection is determined by the amount of inocu-
lum that enters the vascular system of the plant.

The second phase of disease development begins after penetration of the fungus into the vessels and the resulting appearance
of wilt symptoms in the aerial part of the plant. It may begin without the first phase only if the fungus can penetrate the host vascular system and elude the root tissues (for example, through a
wound in the roots caused by nematodes or cultivation, etc.). Hav-
ing penetrated the vascular system, the fungus causes the respon-

sive defense reactions in the vessels. Such a reaction is expressed in necrosis of the cells around the vessels adjoining the point
of infection and also in the plugging of the vessel lumen with gels
and tyloses.

Kling [1938] and Sukhorukov [1942, 1963], on the basis of
their investigations, first pointed out tylosis formation and plugging of the vessels with wound resins as defense reactions of a
plant against infection. Further, a number of workers showed that
plugging, if it occurs quickly and intensively, can localize the infection. The whole matter depends on the intensity with which the
plugging proceeds. The ability to form tyloses occurs in both susceptible and resistant varieties but with different intensity. Thus,
in a resistant variety of tomatoes tyloses appeared 10 days after
infection with *Verticillium albo-atrum* but in a susceptible variety
only after 21 days [Sanha and Wood, 1967]. Differences in rate of
tylosis formation in varieties differing in resistance were seen in
a study of Fusarium wilt of bananas. In the resistant plant metabolites secreted by the fungus damaged mainly the cells adjoining the
point of infection. The possibility of damage by these metabolites
to cells located a short distance from the infection is very slight,
since the transpiration stream stops 24 hr after infection [Beck-
man and Halmos, 1962]. Injury to the cells is shown by browning
and absence of tyloses in the region of severe infection. Above the
point of infection a gradual increase in size and number of tyloses
and in rate of development was seen. In a resistant plant the vessels are fully plugged in 4 days and in a susceptible plant in 10-12
days, during which time the conidia of the fungus are able to spread
throughout the plant.

But localization of the infection by vessel plugging still does
not denote death of the parasite, and therefore reliable protection
of the plant from infection can hardly be assured by means of this
alone. It is natural to assume the emergence of additional protective reactions in the vascular tissues, for example, development
of antibiotic substances of the nature of phytoalexins [Schnathorst
et al., 1966; Sanha and Wood, 1967]. Thus, the entrance of conidia
of *Verticillium albo-atrum* into the xylem vessels of the cotton
stem in 24-78 hr caused an appreciable accumulation of ether-
soluble phenolic compounds, among which gossypol predominated.
Purified gossypol inhibited germination of the conidia at concen-

trations of 50–250 μM (ED$_{50}$) and resembled the described phyto-alexins in its action [Bell, 1967].

The above indicates that plant varieties differing in the degree of their wilt resistance can be distinguished in character and intensity of defense reactions originating either in the roots, the vessels, or both. In order to understand the nature of these reactions and their final outcome, it is necessary, of course, to know what metabolites of the fungus parasite and the host plant interact.

We have already noted the inadequate study of toxins and the failure of attempts to explain the entire course of infectious disease by their effect alone. Nevertheless, proceeding from all that has been said regarding the development of disease and of defense reactions, such as the complex interaction of different metabolites of fungus and plant and the plugging of vessels as one of the defense reactions, some of the available data on toxins might be explained otherwise.

A large number of observations have shown that fusaric acid is a vivotoxin; it has been seen in various plants only 48 hr after their infection, i.e., when the fungus has just begun to grow [Lakshminarayanan and Subramanian, 1955; Kalyanasundaram and Venkata Ram, 1956; Kern and Kluepfel, 1956; Kluepfel, 1957; Page, 1959]. Still, fusaric acid has been observed in a diseased plant at concentrations that do not cause the changes (external and internal) characteristic for a diseased plant, and this furnishes evidence on the basis of which some authors question its role in disease development [Sanwal and Waygood, 1961; Kuo and Scheffer, 1964].

This objection can be removed by considering that in the vessels of a damaged plant, as a result of the plugging that takes place, high concentrations of toxin may appear in isolated locations. On the other hand, different effects of fusaric acid at different concentrations may be the cause of the different changes occurring in resistant and susceptible varieties, which serve to distinguish these varieties from each other. According to the data of Braun [1960], fusaric acid is rapidly detoxified by the plant tissues. The role of fusaric acid in the course of the disease can be assumed on the basis of its pleotropic effect: inhibition of growth, respiration, and polyphenoloxidase; change in permeability of the protoplasm; uncoupling of respiration and oxidative phosphorylation; and necrosis of the parenchyma. It may also enter into synergistic action with other metabolites of the fungus and the plant [Gäumann, 1958]; even at a concentration of 10^{-6} M fusaric acid causes jelling of the perforated membranes in vessels of banana roots [Beckman, 1964]. The observations of Kiessig and Hallier-Kiessig [1957] show that *Verticillium albo-atrum* secretes a toxic substance reminiscent of fusaric acid. As already noted, one of the toxic substances isolated by Akhvlediani [1964, 1966] from the citrus pathogen malsecco caused plugging of the vessels with gel-like substances, products of the life activities of the host cells.

Consequently, the formation of gel plugs in the vessels is possible by means of toxins or auxins without the participation of the pectolytic enzymes of the parasite. The pectic enzymes play a role in the destruction of the plugs rather than in their formation. To explain the role of these enzymes in disease development, investigations of the composition of the plugging substances and of their formation and destruction in a diseased plant are required.

Many investigators point out the pectic nature of the substances plugging the vessels. Beckman et al. [1962] described the appearance of gels in banana infected with *Fusarium oxysporum* and *Fusarium cubense* at the point of infection after 2-5 days. The development of gels was observed both in resistant and in susceptible plants. But in the susceptible plant they had already liquefied on the second day. Observations of the dynamics of formation of these plugs made it possible to determine that the plugging gels are swollen perforated membranes of the vessels and of the primary cell walls. Blue staining by methylene blue, orange by safranin O, rose by hydroxylamine and iron chloride, and red by ruthenium red showed that the gel consists principally of an esterified acid polymer that is pectic in nature. The position of the gel on the upper rather than the lower side of the perforated membrane leads to the conclusion that its formation occurs most probably without participation of the pectic enzymes of the fungus pathogen. There is evidence that pectolytic enzymes, in joint action with other agents, can dissolve these gels [Beckman and Zaroogian, 1967].

Pectic enzymes can thus take part in the destruction of the plugs and thereby overcome a unique defense reaction of the plant. It is possible also that the pectolytic enzymes participate in the first stages of disease development upon penetration of the fungus into the plant.

A number of authors connect the phenomenon of plant resistance to wilt with the ability of the tissues to reduce the activity of the pectolytic enzymes by oxidation products of polyphenolic compounds formed by polyphenoloxidases [Deese and Stahman, 1962a, 1962b; Patil and Dimond, 1967; Patil et al., 1966]. Reduction of the activity of pectolytic enzymes of *Fusarium oxysporum* f. *lycopersici* by different polyphenolic compounds and by extracts of plant tissues was shown by Grossman [1962a, 1962b, 1962c].

Despite the fact that such great importance is attributed to this group of enzymes in disease development, very little is known about their nature, their composition in different pathogens, their properties, their interactions with other metabolites, and their production in the diseased plant. In phytopathological literature little attention has been given, up to the present, to the characterization of enzymes: the character of the disruption of the bond with the pectic molecule, locus of the break, substrate preference, optimal pH, relation to cations, etc. A determination of these characteristics of the enzymes made it possible for us to establish the presence, in the composition of the exoenzymes of the wilt pathogens *Verticillium dahliae* and *Fusarium oxysporum,* of enzymes that cause trans-eliminative breakdown of the pectin molecules [Vasil'-eva and Metlitskii, 1968]. We followed the change in activity of pectin-trans-eliminase (PTE) in the culture liquid in proportion to the fungus growth. The first determinations were performed on the fourth day after inoculation of the medium, and in this period the highest activity was noted, as converted to unit volume of medium and milligram of air-dry mycelium (Fig. 7). Trans-eliminative activity was observed also on the second day after conidial germination.

Enzymes causing trans-eliminative breakdown of pectic substances are distinguished by their properties from polygalacturonases, which are also present in the exoenzymes of the wilt pathogen. The optimal activity of these two enzymes appears at different pH values; pectin-trans-eliminase of *Verticillium dahliae*, according to our observations, is stimulated by calcium ions, whereas polygalacturonase is inhibited by them.

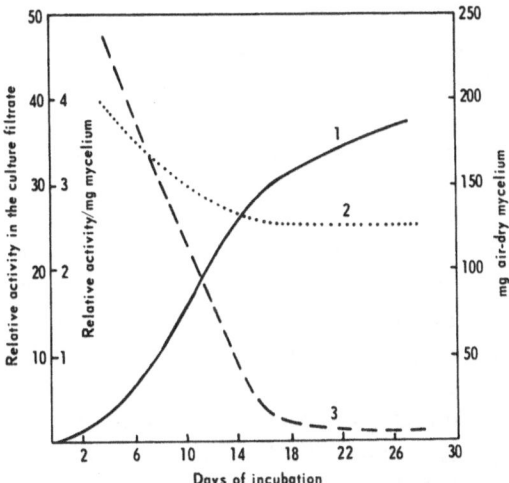

Fig. 7. Change in PTE activity in the culture filtrate in proportion to growth of the fungus *Verticillium dahliae* on nutrient medium. 1) Air-dry weight of mycelium in mg; 2) relative PTE activity; 3) relative PTE activity converted to mg of air-dry weight.

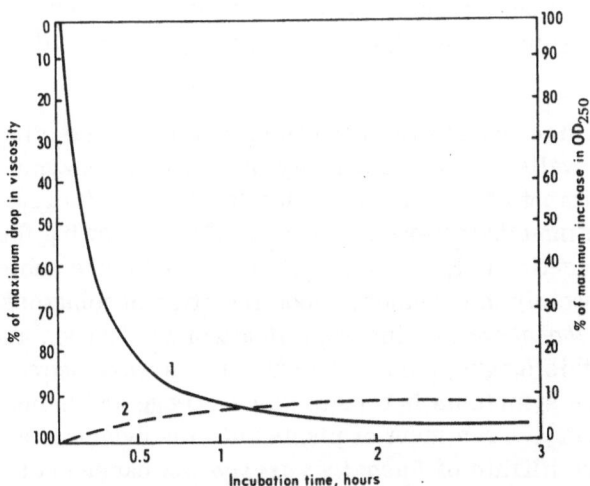

Fig. 8. Character of PTE according to position of the bond rupture in the pectic molecule. 1) Viscometric determination of PTE activity; 2) spectrophotometric determination of PTE activity.

It is well known that enzymes participate basically in the
phenomena of maceration and destruction of pectic gels, where
they behave like endopolygalacturonases and endopectic-trans-
eliminases. In order to judge the type of action of the pectin-
trans-eliminase of *Verticillium dahliae,* we compared viscometric
with spectrophotometric determinations. Actually, viscometric
observations cannot answer the question as to number of developing pectic fragments. This question can be decided only on determination of the freed reducing groups, or spectrophotometrically,
by the increase in double bonds. With the exclusive use of the
above methods, however, it could not be judged whether the break
in the pectic chain occurs at its center or on the ends. Comparison of the viscometric and spectrophotometric determinations established that the pectin-trans-eliminases of *Verticillium dahliae*
behave like endopectin-trans-eliminase (Fig. 8).

Pectin-trans-eliminases are still inadequately studied. Data
are lacking on the action of polyphenols and their oxidation products on these enzymes. It has been shown that indolylacetic acid
can inhibit PTE [Albersheim et al., 1967].

Thus, in the composition of the parasitic exoenzymes are enzymes that affect pectic substances and are distinguished by mode
of action, substrate specificity, optimal pH, relation to other
metabolites, etc.

Of great interest are the results of studies on the effect of
auxins on cell walls and on the activity of pectolytic enzymes. For
example, treatment of a plant with indolylacetic acid increased the
activity of pectinmethylesterase [Wood, 1959]. Following the action of indolylacetic acid, the cell walls appeared more subject to
the action of pectolytic enzymes. Upon infection of tomatoes with
Verticillium albo-atrum, an increase in auxin was shown, and on
cultivation of this fungus, auxin-like substances were secreted
causing disease symptoms in tomato shoots [Pegg and Selman,
1959; Pegg, 1965]. Treatment of plants with indolylacetic acid or
with the culture filtrate of *Fusarium oxysporum* caused collapse
of the vessels because of destruction of the processes of second-
ary thickening of the cell walls. A similar collapse of vessels was
seen in plants infected with wilt [Chambers and Corden, 1963].

Mace and Solit [1966] showed that indolylacetic acid causes the formation of tyloses in banana roots, but they also noted that in a diseased plant it is very difficult to tell whether the auxins belong to the tissues of the host plant or to the fungus parasite.

Some authors believe that certain compounds formed on interaction of plant and fungus parasite are necessary for the formation of tyloses. Characteristically, tylosis formation is lacking at points of severe infection. Tyloses increase in number and size with distance from the place of infection.

Sukhorukov [1942, 1963] stated the hypothesis that cell substances stimulating the wound reaction are obtained from neighboring cells surrounding the infected vessel. This substance may be of the nature of traumatin [Sadasivan, 1961].

The interaction of phenols and their oxidation products with pectolytic enzymes, tylosis formation, and auxins has been considered in a number of works [Farkas and Kiraly, 1962; Sanha and Wood, 1967; etc.]. In the preceding chapter, we have already noted the ability of polyphenol oxidation products to suppress the activity of pectolytic enzymes. Similar results were obtained in experiments with pectolytic enzymes of wilt pathogens. It is even suggested that phenols and their oxidation products, in combination with gel, form more resistant plugs through which the fungus is unable to penetrate. The same applies to tyloses, which are generally lacking in the browned vessels [Beckman, 1966].

The ability of some phenolic compounds to fluoresce in ultraviolet light and the change in this ability with their conversion are well known. Fluorescence of plant tissues is described as one of the earliest symptoms of Fusarium wilt in cotton [Subba-Rao, 1954]. The nature of the substances causing fluorescence has not yet been determined; it is proposed that they are degradation products of the host tissues [Sadasivan, 1961]. They may, however, also belong to the parasite.

Although the nature of the substances causing fluorescence is unknown, the established correlation between rate of advance of the front of fluorescing substances in the stem and the degree of resistance of a cotton variety to *Verticillium dahliae* is of great interest [Guseva, 1965; Ozeretskovskaya and Guseva, 1966]. On

Fig. 9. Hypothetical scheme of the interaction of host plant and fungus. 1) Defense reactions in the cortex (suberization, lignification); 2) defense reactions in the vascular tissues; 3) inhibition of conidial germination; 4) formation of auxiliary xylem rings.

the second day after infection of the plant, this front advanced extremely slowly in resistant, as compared with susceptible, varieties. It is very important that such differences are observed on infection of plants even in the cotyledon stage, since it suggests the possibility of using the character of tissue fluorescence as a diagnostic method for plant susceptibility in the practice of selection.

We shall now attempt, on the basis of these results, to present, if only schematically, the most important of the presently described defense reactions of plants to the introduction of a wilt pathogen, in the aggregate (Fig. 9).

Upon penetration of a fungus from the soil into the roots, the first responsive defense reactions are suberization, lignification, and deposit of gum in cell walls and root tissues. These reactions, with definite conditions of the external environment and fungus − plant combinations (different aggressiveness of parasite and resistance of host), lead either to localization of the infection or to its exclusion. The fungus advances through the cortex tissues intercellularly or intracellularly, either by means of exoenzymes or without their participation. The advance through cortex tissues proceeds by means of hyphal growth. A defensive role of the polyphenol − polyphenoloxidase system, involving the inhibition of exoenzymes and suppression of hyphal growth, may occur also in this stage.

After penetration into the vascular system of the plant, the fungus forms conidia that enter the transpiration stream. Here the fungus excretes substances that induce formation of gel deposits and also swelling of the cell walls and of the septa separating the individual cells of the vessels. The substances inducing these reactions may be toxins and growth substances of the fungus. The conidia settle in the gel and begin to germinate. The pectolytic (possibly also proteolytic, cellulolytic, and hemicellulolytic) enzymes formed by them destroy the gel and the swollen septa of the cells, permitting distribution of the newly formed conidia higher through the plant. In a resistant plant, however, either the gel is strong enough or the exoenzymes are actually inactivated by polyphenols that are freed from glycosides by the action of fungal β-glucosidase or the oxidation of the polyphenoloxidase of the plant. The parenchyma cells surrounding the infected vessel and contiguous to the fungus die (necrotize) as a result of destruction of their oxidative processes. The "necrohormones" that form, and possibly also auxins secreted by the fungus and developing in the host tissues, stimulate growth of the higher-placed cells and their projection into the vessel cavity. The tyloses formed in this way plug the vessel and localize the infection.

Finally, defense reactions also include the formation of substances of the nature of phytoalexins, which inhibit growth of the fungus and conidial germination. Their formation may be induced by exoenzymes of the fungus. With intensive development of the defense reaction by vessel plugging, the plant may survive and provide new xylem rings, which phenomenon is associated with a change in level of the regulators of growth and development in the infected plant brought about by the infection. These interrelationships are subjected to the influence of both internal and external factors.

The outcome of the interactions depends on how quickly and in what quantities metabolites are formed by the fungus responsible for the infection and how rapidly and intensively the host reactions begin.

Although the ideas presented are based on numerous experimental data and observations, they include much that is still unclear and hypothetical. Only further investigations will show how authentic the suggested system may be.

Conclusion

The entire presentation leads us to the conclusion that in recent years, thanks to the efforts of many groups of scientists, definite progress has been made in the study of the biochemical nature of plant resistance to phytopathogenic microorganisms. A
step forward has been taken toward unraveling one of the inner-
most secrets of animate nature: by what method higher plants,
many times exceeded by microorganisms in rate of multiplication,
plasticity, variability, and adaptation to unfavorable conditions of
existence, are rendered not completely defenseless to parasites
so that their death is more the exception than the rule.

No doubt now remains that plant resistance to many diseases
is connected with the presence of antibiotic substances in plant tissues and with plants' ability to develop such substances in response
to infection. Since these substances are metabolites inherent in
each plant species, they are found both in resistant and in susceptible varieties of that species. The differences between varieties involve the quantity of antibiotic substances contained in intact tissues, perhaps to a greater extent the intensity of their generation in response to infection, and finally the character of subse-
quent conversions which may lead to a marked increase in antibiotic activity. With the formation and conversion of antibiotic substances are associated defense reactions, especially reactions to
wounding and the necrotic reaction, that are widely distributed in
the plant world.

The various antibiotic substances of plants are sharply distinguished from each other by their activity, and therefore different roles are assigned to them in the phenomena of phytoimmunity. It is thus entirely probable that resistance to a definite disease is determined not by one substance, however active it may be, but by several that act synergistically.

Single antibiotic substances of plants affect the most diverse pathogens and originate in plants in response to all kinds of disturbances. And, further, there is ground to consider that substances which inhibit or stimulate growth of the parasite exert exactly the same effect on growth processes of the plant itself. This phenomenon, we believe, is one of the manifestations of the ability of animate nature to use the same mechanisms for different purposes.

Although a very important role in plant defense reactions is assigned to antibiotic substances, it is impossible to explain, solely by their formation, the mechanism of interrelationships emerging on contact of plant and parasite, the outcome of which will determine whether the plant remains healthy or dies. In the tracheomycotic diseases, as an example, we tried to show that resistance to a definite disease depends on many factors and that, in the course of disease development, along with the formation of antibiotic substances other protective reactions, suberization, lignification, and formation of gel-like substances, auxins, tyloses, and new xylem vessels also play a part.

The emergence of all these reactions is dependent not only on a change in the activity of enzymes existing in intact tissue, but also on the formation of their isozymes and reorganization of respiratory gas exchange, the source of energy, and supplier of many intermediate compounds for biosynthetic processes.

Phytoimmunity is thus based on a multiplicity of defense reactions involving extremely varied substances.

"Contemporary immunological theories," wrote Burnet [1962], "for the most part are too simple, too mechanistic in form to be biologically probable." And, although this was said more than five years ago in regard to the mechanism of immunity in vertebrates, which are more complex than invertebrates and plants, the same statement applies to a considerable extent also to contemporary theories of phytoimmunity.

Periodic reviews of various investigations in the light of current knowledge seem of great importance in this context. Generalizations based on such reviews make it possible to structure new experiments capable of proving or amplifying earlier ones. We have striven for this in the composition of the present book. We have not avoided statements of opinion, even in those cases where experimental data are still inadequate. This has been done with the single purpose of stimulating further experiments aimed at the development of a fruitful theory of the immunity of plants.

All critical comments will be received with deep appreciation.

References

Akazawa, T. 1956. Science, 123, 1075.

Akazawa, T., and Uritani, I. 1961. Phytopathology, 51, No. 10.

Akhvlediani, K. S. 1964. Candidate's Dissertation, Tbilisi.

Akhvlediani, K. S. 1966. In: Biochemical Principles of Plant Protection. Moscow, Nauka, p. 42.

Albersheim, P., Nevins, D. J., English, P. D., and McNab, J. M. 1967. Abstracts of papers at the International Symposium on Plant Biochemical Regulation in Viral and Other Diseases or Injury, August 17-19, 1967.

Allen, P. J. 1966. Ann. review, 56, No. 3, 255.

Bakh, A. N. 1950. Collected Works on Chemistry and Biochemistry. Moscow, Izd. AN SSSR.

Beauverie, M. M. 1901. Compt. Rend. Acad. Sci., 133, 107.

Beckman, C. H. 1964. Ann. Rev. Phytopathology, No. 2, 231.

Beckman, C. H. 1966. Phytopathology, 56, No. 7, 821.

Beckman, C. H., and Halmos, S. 1962. Phytopathology, 52, 893.

Beckman, C. H., and Zaroogian, G. E. 1967. Phytopathology, 57, No. 1, II.

Beckman, C. H., Halmos, S., and Mace, M. E. 1962. Phytopathology, 52, 134.

Becker, Z. E. 1963. Physiology of Fungi and Their Practical Utilization. Izd. MGU.

Bell, A. A. 1967. Phytopathology, 57, No. 7, 759.

Bonner, J., and English, J., Jr. 1938. Plant Physiol. 13, No. 2, 331.

Braun, E. C., and Pringle, R. B. 1959. In: Plant Pathology Problems and Progress 1908-1958. Madison, University of Wisconsin Press.

Braun, R. 1960. Phytopathol. Z., 39, 197.

Burnet, F. M. 1962. The Integrity of the Body — A Discussion of Modern Immunological Ideas. Cambridge, Harvard University Press.

Calo, N., Marks, J., and Varner, J. E. 1957. Nature, 180, No. 1142.

Chambers, H. L., and Corden, M. E. 1963. Phytopathology, 53, 1006.

Chester, K. S. 1933. Quart. Rev. Biol., 8, 129, 275.

Clerici, E., Guidotti, G., Sambo, G., and Bazzano, E. 1960. Proc. Soc. Exptl. Biol. and Med., 105, 377.

103

Comes, O. H. 1914. Ann. Scuola Sup. Agr. Portici, Ser. II, 12.

Condon, P., and Kuč, J. 1960. Phytopathology, 50, 267.

Condon, P., Kuč, J., and Draudt, H. 1963. Phytopathology, 53, 1244.

Cruickshank, J. A. M. 1961. Austr. J. Biochem. Sci., 15, No. 1, 147.

Cruickshank, J. A. M. 1963. Annual Review of Phytopathology, 1, 351.

Cruickshank, J. A. M. 1965. In: "Ecology of Soil-borne Plant Pathogens." Prelude to Biological Control. 325.

Cruickshank, J. A. M., and Manruk, M. 1960. J. Austr. Inst. Agr. Sci., 26, 369.

Cruickshank, J. A. M., and Perrin, D. R. 1960. Nature, 187, No. 4739, 799.

Cruickshank, J. A. M., and Perrin, D. R. 1961. Austr. J. Biol. Sci., 14, 336.

Cruickshank, J. A. M., and Perrin, D. R. 1962. Austr. J. Biol. Sci., 16, No. 1.

Cruickshank, J. A. M., and Perrin, D. R. 1963. Life Sci., 680.

Cruickshank, J. A. M., and Perrin, D. R. 1965. Austr. J. Biol. Sci., 18, 828.

Cutter, V. M. 1951. Trans. N. Y. Acad. Sci. Ser. 2, 14, No. 2, 103.

Daly, J. M., and Inman, R. E. 1958. Phytopathology, 48, 91.

Davis, D. 1965. Phytopathology, 55, No. 10.

Deese, D. C., and Stahmann, M. A. 1962a. Phytopathology, 52, 255.

Deese, D. C., and Stahmann, M. A. 1962b. Phytopathol. Z., 46, No. 1, 53.

Dimond, A. E. 1955. Ann. Rev. Plant Physiol., 6, 329.

Dimond, A. E., and Waggoner, P. E. 1953. Phytopathology, 43, 229.

Drobot'ko, V. G., Aizenman, B. E., Shvaiger, M. O., Zelenukha, S. I., and Mandrika, T. P. 1958. Antimicrobial Substances in Higher Plants. Kiev, Izd. AN SSSR.

Dubinin, N. P., and Glembotskii, Ya. L. 1967. Population Genetics and Selection. Moscow, Izd. Nauka.

D'yakov, Yu. T. 1965. Tezisy Dokladov IV Bsesoyuznogo Soveshchaniya po Immunitetu Sel'skokh. Rastenii. Kishinev.

D'yakov, Yu. T., and Kogan, I. G. 1966. Sel'khoz. Biologiya, 1, No. 5.

Dzhaparidze, L. I., and Kanchaveli, N. Z. 1948. Soobshcheniya AN Gruzinskoi SSR, 9, No. 7. Tbilisi.

Eglits, M. 1933. Phytopathol. Z., 5, No. 4, 343.

English, J., Jr., Bonner, J., and Haagen-Smit, A. J. 1939. J. Am. Chem. Soc., 61, No. 12, 3434.

Ernster, L. 1957. Biochem. J., 65, 44.

Farkas, G., and Kiraly, Z. 1962. Phytopathol. Z., 44, 2, 8.

Farkas, G., and Ledingham, G. 1959. Canad. J. Microbiol., 5, 37.

Farkas, G., Dezsi, L., Horvotu, M., Kisban, K., and Udvanly, J. 1964. Phytopathol. Z., 49, 4.

Fischer, H. 1950. Phytopathol. Z., 16, 171.

Fischer, K. 1965. Phytopathology, 55, No. 4.

Flor, H. H. 1956. Adv. Genetic, 8, 29.

Gallegly, M. E., and Niederhauser, J. S. 1959. In: Plant Pathology Problems and Progress 1908-1958. Madison, University of Wisconsin Press.

Garber, R. H., and Houston, B. R. 1966. Phytopathology, 56, 10, 1121.

Gäumann, E. 1951. Adv. Enzymol., 2, 401.

Gäumann, E. 1954. Infectious Diseases of Plants [Russian translation of "Pflanzliche Infektionslehre"]. Moscow, IL.

Gäumann, E. 1957. Phytopathology, 47, 342.

Gäumann, E. 1958. Phytopathology, 48, 670.

Gäumann, E. 1963. Compt. Rend., 257, 2372.

Gäumann, E. 1963/1964. Phytopathol. Z., 49, 211.

Gäumann, E., and Hohl, H. R. 1960. Phytopathol. Z., 38, 93.

Gäumann, E., and Kern, H. 1959. Phytopathol. Z., 36, 1.

Gäumann, E., Nuesch, J., and Rimpau, R. H. 1960. Phytopathol. Z., 38, 274.

Gorlenko, M. V. 1962. Short Course in Plant Immunity to Infectious Diseases. Moscow, Sovetskaya Nauka.

Gothoskar, S. S., Scheffer, R. P., Walker, J. C., and Stahmann, M. A. 1953. Phytopathology, 43, 535.

Gothoskar, S. S., Scheffer, R. P., Walker, J. C., and Stahmann, M. A. 1955. Phytopathology, 45, 381.

Gottlieb, D. 1943. Phytopathology, 33, 1111.

Gottlieb, D., and Garner, J. M. 1946. Phytopathology, 36, No. 7, 557.

Grossmann, F. 1962a. Phytopathol. Z., 44, 361.

Grossmann, F. 1962b. Phytopathol. Z., 45, 1.

Grossmann, F. 1962c. Phytopathol. Z., 45, 139.

Guseva, N. N. 1965. In: Methodological Instructions in Diagnosis of Verticillium Wilt of Cotton. Leningrad.

Haberland, G. 1922. Biol. Zentralblatt, 145.

Hackett, D., Haas, D., Griffiths, S., and Niederpruem, D. 1959. Plant Physiol., 31, No. 8.

Hampton, R. E. 1962. Phytopathology, 52, 413.

Heitefuss, R., Buchanan-Davidson, D. J., Stahmann, M. A., and Walker, J. C. 1960. Phytopathology 50, No. 3, 198.

Hietala, P. K. 1960. Ann. Acad. Sci. Fennicae Ser. A2, 100, 1.

Hiroe, J., and Nishimura, S. 1956a. J. Agr. Chem. Soc. Japan, 30, 528.

Hiroe, J., and Nishimura, S. 1956b. Phytopathol. Soc. Japan, 20, 161.

Hiura, M. 1943. Rept. Gifu Agric. College, 50, 1.

Holowczak, J., Kuč, J., and Williams, E. B. 1962. Phytopathology, 52, 699.

Hughes, J. C., and Swain, T. 1960. Phytopathology, 50, 398.

Husain, A., and Dimond, A. E. 1960. Phytopathology, 50, No. 5, 329.

Husain, A., and Kelman, A. 1959. Plant Pathology, 1, 144. New York, Academic Press.

Imaseki, H., and Uritani, I. 1964. Plant Cell Physiol., 5, 133.

Jermyn, M. A., and Thomas, R. 1954. Biochem. J., 56, No. 4, 631.

Johnson, G., and Schaal, L. 1957. Am. Potato J., 34, No. 7, 200.

Kalyanasundaram Rand, C. S., and Venkata Ram. 1956. J. Indian Bot. Soc., 35, 7.

Kamal, M., and Wood, R. K. S. 1956. Ann. Appl. Biol., 44, No. 2, 322.

Kawashima, N., and Uritani, J. 1963. Agr. Biol. Chem., 27, 409.

Kern, N., and Kluepfel, D. 1956. Experientia, 12, 181.

Kiessig, R., and Hallier-Kiessig, R. 1957. Phytopathol. Z., 31, 185.

Kirkham, D. S. 1957. J. Gen. Microbiol., 17, 491.

Kling, E. G. 1938. Proceedings (Trudy) of the K. A. Timiryazev Institute of Plant Physiology, Vol. 11, No. 1. Moscow, Izd. AN SSSR.

Kluepfel, D. 1957. Phytopathol. Z., 29, 349.

Korableva, N. P., and Potapova, L. M. 1966. In: Biochemicâl Principles of Plant
     Protection. Moscow, Nauka.
Kovalenok, A. 1944. Effect of Phytoncides on Protozoa. "Phytoncides," Tomsk.
Kozo-Polyanskii, B. M. 1946. "What Are Phytoncides?" Nauka i Zhizn'.
Krasil'nikov, N. A., Khodzhibaeva, S. M., and Mirchink, G. T. 1965. Agrokhimiya,
     No. 10.
Krasil'nikov, N. A., Khodzhibaeva, S. M., and Mirchink, G. T. 1966. Sel'khoz.
     Biologiya, 1, No. 1, 107.
Kubota, T., and Matsurua, T. 1953. J. Chem. Soc. Japan. Pure Chem. Sect., 74, 248.
Kuč, J. 1966. Ann. Rev. Microbiol., 20, 337.
Kuč, J., Williams, E. B., and Shay, J. R. 1957. Phytopathology, 47, 21.
Kuo, M. S., and Scheffer, R. P. 1964. Phytopathology, 56, No. 9, 1041.
Kuprevich, V. F. 1947. Physiology of the Diseased Plant. Moscow-Leningrad, Izd.
     AN SSSR.
Kursanov, A. L. 1943. Biokhimiya, 8, Nos. 2-3, 109.
Lacey, B. W. 1958. In: The Strategy of Chemotherapy, S. T. Cowan and E. Rowatt,
     eds. Cambridge, Society for General Microbiology.
Lakshminarayanan, K., and Subramanian, D. 1955. Nature, 176, 697.
Lampitt, L., Bushill, J., Rooke, H., and Jackson, E. 1943. J. Soc. Chem. Ind., 62, 48.
Lee, S. G., and Chasson, R. M. 1966. Physiol. plantarum, No. 19, 199.
Linderberg, G. 1949. Svensk. Bot. Tidskr., 43, 438.
Lipsits, D. V. 1965. Doctoral Dissertation. Moscow.
Lovrekovich, L., and Farkas, G. 1965. Nature, 205, 823.
Ludwig, R. U. 1952. Techn. Bull., 20, 385.
Ludwig, R. A. 1960. Plant Pathology, 2, 315.
Mace, M. E., and Solit, E. 1966. Phytopathology, 56, No. 2.
Malcolmson, J. P. 1965. Europ. Potato J., 8, 183.
Massee, G. 1905. Phil. Trans. Roy. Soc. Lond., 197, 7.
McDonnell, K. 1958. Nature, 182, 1025.
McDonnell, K. 1962. Trans. Brit. Mycol. Soc., 45, No. 1, 55.
McIntyre, G. A. 1965. Phytopathology, 55, No. 10.
McKee, R. K. 1954. Ann. Appl. Biol., 41, 417.
McKee, R. K. 1955. Ann. Appl. Biol., 43, 147.
Metlitskii, L. V., and Akhvlediani, K. S. 1965. Doklady IV Vsesoyuznogo
     Soveshchaniya po Immunitetu Sel'skokh. Rastenii. Kishinev.
Metlitskii, L. V., and Mukhin, E. N. 1964. In: Biochemistry of Fruits and Vegetables
     (Immunity and Dormancy of the Potato, Fruits, and Vegetables). Moscow, Izd.
     Nauka.
Metlitskii, L. V., and Volkind, I. L. 1966. Storage of Potatoes under Conditions of
     Active Ventilation. Moscow, Izd. Ekonomika.
Metlitskii, L. V., Ozeretskovskaya, O. L., and Chalenko, G. I. 1965. Trudy VIZR,
     No. 26.
Metlitskii, L. V., Ozeretskovskaya, O. L., Chalenko, G. I., and Strokova, G. A.
     1965. Doklady Akad. Nauk SSSR, 160, No. 4.
Miyoshi, M. 1894. Bot. Z., 52, 1.
Mizukami, T. 1953. Ann. Phytopathol. Soc. Japan, 17, 57.

Müller, K. O. 1958. Austr. J. Biol. Sci., 2, No. 3, 275.

Müller, K. O., and Börger, H. 1940. Arb. Biol. Reichsanstalt Land.-u.-Forst-wirtsch., Berlin, 23, 183.

Nelson, J. M., and Dawson, C. R. 1944. Adv. Enzymol., 4, 99.

Oku, H., 1960. Phytopathol. Z., 38, 342.

Oparin, A. I., and Kuplenskaya, O. I. 1935. Problems of Immunity of Cultivated Plants. Trudy Maiskoi Sessii AN SSR, Vol. 60.

Ozeretskovskaya, O. L., and Guseva, N. N. 1966. Trudy VNIIZR, No. 26.

Ozeretskovskaya, O. L., and Metlitskii, L. V. 1966. In: Biochemical Principles of Plant Protection. Moscow, Izd. Nauka.

Ozeretskovskaya, O. L., and Vasyukova, N. I. 1965. Doklady Akad. Nauk SSSR 161, No. 4, 968.

Ozeretskovskaya, O. L., Vasyukova, N. I., and Davydova, M. A. 1968a. Prikladnaya Biokhimiya i Mikrobiologiya (in press).

Ozeretskovskaya, O. L., Vasyukova, N. I., and Metlitskii, L. V. 1968b. Doklady Akad. Nauk SSSR (in press).

Page, O. T. 1959. Phytopathology, 49, 230.

Patil, S. S., and Dimond, A. E. 1967. Phytopathology, 57, No. 5, 492.

Patil, S. S., Zucker, M., and Dimond, A. E. 1966. Phytopathology, 56, 971.

Pegg, G. F. 1965. Nature, 208, No. 5016.

Pegg, G. F., and Selman, J. W. 1959. Ann. Appl. Biol., 47, 222.

Perrin, D. R., and Bottomley, W. 1962. J. Am. Chem. Soc., 84, No. 10, 1919.

Perrin, D. R., and Cruickshank, J. A. M. 1965. Austr. J. Biol. Sci., 18, 803.

Person, C. 1959. Canad. J. Bot., 37, 1101.

Petrov, D. F. 1959. Selection of Microbes. Moscow, Medgiz.

Pfeffer, W. 1884. Untersuchungen Bot. Inst. Tübingen, 1, 3, 363.

Pierson, C. F., Gothoskar, S. S., Walker, J. C., and Stahmann, M. A. 1955. Phytopathology, 45, 524.

Presley, J. T., Carns, H. R., Taylor, E. E., and Schnathorst, W. C. 1966. Phyto-pathology, 56, No. 3, 375.

Pringle, R. B., and Scheffer, R. P. 1964. Ann. Rev. Phytopathology, 2, 133.

Prokoshev, S. M. 1943. Biokhimiya, 8, Nos. 2-3, 124.

Ray, M. J. 1901. Compt. Rend. Acad. Sci., 133, 307.

Rubin, B. A., and Aksenova, V. A. 1957. Biokhimiya, 22, Nos. 1-2, 202.

Rubin, B. A., and Aksenova, V. A. 1964. Fiziologiya Rastenii, 11, No. 1, 59.

Rubin, B. A., and Artsikhovskaya, E. V. 1960. Biochemistry and Physiology of Plant Immunity. Moscow, Izd. AN SSSR.

Rubin, B. A., and Ozeretskovskaya, O. L. 1962. In: Biochemistry of Fruits and Vegetables, No. 7, 133. Moscow, Izd. AN SSSR.

Rubin, B. A., Aksenova, V. A., and Ladygina, M. E. 1966. Doklady Akad. Nauk SSSR, 166, No. 5, 1236.

Rudolph, K., and Stahmann, M. A. 1964. Z. f. Pflanzenkrankheiten u. Pflanzenschutz, 71, 107.

Sadasivan, T. S. 1961. Ann. Rev. Plant Physiol., 12, 449.

Sadasivan, T. S., and Kalyanasundaram, R. 1956. Proc. Indian Acad. Sci., 43, 271.

Sal'kova, E. G., and Bekbulatova, P. 1965. Prikladnaya Biokhimiya i Mikrobiologiya, 1, No. 4.

Sal'kova, E. G., and Platonova, T. A. 1967. Tezisy Dokladov na Simpoziume po Fenol'nym Coedineniyam. Moscow.

Sanha, A. K., and Wood, R. K. S. 1967. Ann. Appl. Biol., 59, 143.

Sanwal, B. D., and Waygood, E. G. 1961. Experientia, 17, 174.

Scheffer, R. P., and Walker, J. C. 1954. Phytopathology, 44, 94.

Scheffer, R. P., Gothoskar, S. S., Pierson, C. F., and Collins, R. P. 1956. Phytopathology, 46, 83.

Schellenbaum, M. 1959. Technische Hochschule, Zurich, Prom., 2977.

Schnathorst, W. C., Coplin, D., Presley, J. F., and Carns, H. R. 1966. Phytopathology, 56, No. 8, 871.

Shaw, M., and Hawkins, A. R. 1958. Canad. J. Bot., 36, 1.

Shaw, M., and Samborski, D. J. 1956. Canad. J. Bot., 34, 389.

Shaw, M., Brown, S. A., and Rudd Jones, D. R. 1954. Nature, 173, 768.

Skulachev, V. P. 1962. Relation of Oxidation and Phosphorylation in the Respiratory Chain. Moscow, Izd. AN SSSR.

Sokolova, V. E. 1964. In: Biochemistry of Fruits and Vegetables. Moscow, Izd. AN SSSR, p. 36.

Sokolova, V. E., Zvyagintseva, Yu. V., and Pel'ts, M. Ya. 1967. Doklady Akad. Nauk SSSR, 173, No. 6.

Sondheimer, E. 1961. Phytopathology, 52, 182.

Stahmann, M. A. 1964. Tagungsber. Dtsch. Akad. Landw. Wiss., Berlin.

Stakman, E. C., and Harrar, J. G. 1957. Principles of Plant Pathology. New York, Ronald Press.

Staples, R. C., and Stahmann, M. A. 1963. Science, 140, 1320.

Subba-Rao, N. S. 1954. J. Indian Bot. Soc., 33, 443.

Subramanian, D., and Saraswathi-Devi, L. 1959. Plant Pathology, 9, 313.

Sukhorukov, K. T. 1942. Physiology and Immunity of Plants. Moscow, Izd. AN SSSR.

Sukhorukov, K. T. 1963. In: Scientific Principles of Crop Protection. Moscow, Izd. AN SSSR.

Suzuki, N. 1957. Bull. Nat. Inst. Agr. Sci. Japan, Ser. C., No. 8, 69.

Szent-Györgyi, A., and Cietorisz, K. 1931. Biochem. Z., 233, 236.

Talboys, P. W. 1957. Trans. Brit. Mycol. Soc., 40, No. 3, 415.

Talboys, P. W. 1958a. Trans. Brit. Mycol. Soc., 41, 227.

Talboys, P. W. 1958b. Trans. Brit. Mycol. Soc., 41, No. 2, 249.

Talboys, P. W. 1958c. Trans. Brit. Mycol. Soc., 41, No. 2, 242.

Talboys, P. W. 1964. Nature, 202, No. 4930, 361.

Tokin, B. P. 1942. Bactericides of Plant Origin (Phytoncides). Moscow, Medgiz.

Tokin, B. P. 1948. Phytoncides. Moscow, Izd. AN SSSR.

Tokin, B. P. 1966. Itogi Rabot IV Vsesoyuznogo Soveshchaniya po Immunitety Sel'skokh. Rastenii. Kishinev.

Tomiyama, K. 1955. Ann. Phytopathol. Soc. Japan, 19, 149.

Tomiyama, K. 1956. Ann. Phytopathol. Soc. Japan, 20, 165.

Tomiyama, K. 1957. Ann. Phytopathol. Soc. Japan, 22, 129.

Tomiyama, K., and Stahmann, M. 1964. Plant Physiol., 39, No. 3, 483.

Tomiyama, K., Ishizaka, N., Sato, N., Masamune, T., and Katsui, N. 1967. Abstr. Internat. Symposium on Plant Biochem. Regulation in Viral and Other Diseases.

Uehara, K. 1959. Ann. Phytopathol. Soc. Japan, 24, 224.

Uritani, I. 1966. Arch. Biochem. and Biophys., 2, 113.

Uritani, I., Akazawa, T., and Uritani, M. 1954. Nature, 174, 1060.

Uritani, I., Uritani, M., and Yamada, H. 1960. Phytopathology, 50, 30.

Van den Ende, G. 1958. Acta Bot. Neerlandica, 7, 665.

Vasil'eva, K. V., and Metlitskii, L. V. 1968. Doklady Akad. Nauk SSSR, 178, No. 6.

Vavilov, N. I. 1935. Study of Plant Immunity to Infectious Diseases. Moscow, Sel'khozgiz.

Verderevskii, D. D. 1957. Methodology for Study of Phytoncidal (Antimicrobial) Properties of Plants in Phytopathology. Kishinev.

Verderevskii, D. D. 1959. Immunity of Plants to Parasitic Diseases. Moscow, Sel'khozgiz.

Verderevskii, D. D., Kuporitskaya, K. I., Zastenchik, N. I., Gatina, E. Sh., Naidenova, I. N., Ves'min'sh, G. E., and Verderevskaya, T. D. 1964. In: Questions of Immunity and Sanitation of Plants. Izd. Urozhai, Kiev, p. 83.

Waggoner, P. E., and Dimond, A. E. 1955. Phytopathology, 45, 79.

Weber, D. J., and Stahmann, M. A. 1966. Phytopathology, 56, No. 9.

Weintraub, M., Ragetli, H. W., and Dwurazno, M. M. 1964. Canad. J. Bot., 42, No. 5, 541.

Wheeler, H., and Luke, H. H. 1963. Ann. Rev. Microbiol., 17, 223.

Winstead, N. N., and Walker, J. C. 1954a. Phytopathology, 44, 153.

Winstead, N. N., and Walker, J. C. 1954b. Phytopathology, 44, 159.

Wood, R. K. 1959. In: Plant Pathology Problems and Progress 1908-1958. Madison, University of Wisconsin-Press.

Wood, R. K. 1961. Ann. Rev. Plant Physiol., 12, 299.

Wu, L. C., and Scheffer, R. P. 1962. Phytopathology, 52, 354.

Yarwood, C. E. 1953. Phytopathology, 43, No. 12, 675.

Yarwood, C. E., and Jacobson, L. 1955. Phytopathology, 45, 43.

Zhukovskii, P. M. 1966. Itogi Raboty IV Vsesoyuznogo Soveshchaniya po Immunitetu Sel'skokh. Rastenii. Kishinev.

Zdrodovskii, P. F. 1961. Problems of Infection and Immunity. Moscow, Medgiz.

Index

adenosine diphosphoric acid 39, 40, 63
adenosine triphosphoric acid 38-41, 45, 63
alliin 10
Anacamptis pyramidalis 33
antibiosis 7
Aphanomyces eutriches 29
apotomic oxidation 69
apple
    infected with *Botrytis cinerea* 47
    resistance to *Venturia inaequalis* 29
Ascochyta pisi 25, 27
auxins 91, 94-95, 97
Bacillus phytophthorus 47
banana
    effect of fusaric acid on 91
    effect of indolylacetic acid on 95
    infected with *Fusarium oxysporum* 91
    wilt of 89
bean
    infected with *Pseudomonas phaseolicola* 15
    infected with rust 47
    infected with *Sclerotinia fructicola* 76
    infected with *Uromyces phaseoli* 15
    interaction with *Monilia fructicola* 23
    phaseollin from 33, 34
    wound periderm in 54
beet roots
    antibiosis 7
    resistance to storage rot 10
    black rot of sweet potato 68-69

beet roots (cont)
    Botrytis allii 10
    Botrytis cinerea 43, 46, 47, 51, 68, 73
cabbage 15
    infected with *Botrytis cinerea* 43, 46
caffeic acid
    effect on hyphal growth 70-72
    effect on mitochondria 74
    effect on polygalacturonase activity 73
    effect on wound periderm 56-58, 62, 64-65
    fungitoxic activity 64-65, 70-71
carrot
    interaction with *Ceratocystis fimbriata* 23
    isocoumarine in 24, 25, 28
catechol 72, 73
Ceratocystis fimbriata 23
Ceratocystis ulmi 24
Ceratostomella fimbriata 15, 24, 30-32, 45, 75
chakonin 62, 71
chemotropic theory of plant resistance 6, 7
chlorogenic acid 8
    effect on hyphal growth 70-72
    effect on mitochondria 74
    effect on polygalacturonase activity 73
    effect on wound periderm 56-58, 64-65
    fungitoxic activity 70

111